This series aims at speedy, informal, and high level information on new developments in mathematical research and teaching. Considered for publication are:

1. Preliminary drafts of original papers and monographs

2. Special lectures on a new field, or a classical field from a new point of view

3. Seminar reports

4. Reports from meetings

Out of print manuscripts satisfying the above characterization may also be considered, if they continue to be in demand.

The timeliness of a manuscript is more important than its form, which may be unfinished and preliminary. In certain instances, therefore, proofs may only be outlined, or results may be presented which have been or will also be published elsewhere.

The publication of the "*Lecture Notes*" Series is intended as a service, in that a commercial publisher, Springer-Verlag, makes house publications of mathematical institutes available to mathematicians on an international scale. By advertising them in scientific journals, listing them in catalogs, further by copyrighting and by sending out review copies, an adequate documentation in scientific libraries is made possible.

Manuscripts

Since manuscripts will be reproduced photomechanically, they must be written in clean typewriting. Handwritten formulae are to be filled in with indelible black or red ink. Any corrections should be typed on a separate sheet in the same size and spacing as the manuscript. All corresponding numerals in the text and on the correction sheet should be marked in pencil. Springer-Verlag will then take care of inserting the corrections in their proper places. Should a manuscript or parts thereof have to be retyped, an appropriate indemnification will be paid to the author upon publication of his volume. The authors receive 25 free copies.

Manuscripts in English, German or French should be sent to Prof. Dr. A. Dold, Mathematisches Institut der Universität Heidelberg, Tiergartenstraße or Prof. Dr. B. Eckmann, Eidgenössische Technische Hochschule, Zürich.

Die „*Lecture Notes*" sollen rasch und informell, aber auf hohem Niveau, über neue Entwicklungen der mathematischen Forschung und Lehre berichten. Zur Veröffentlichung kommen:

1. Vorläufige Fassungen von Originalarbeiten und Monographien.

2. Spezielle Vorlesungen über ein neues Gebiet oder ein klassisches Gebiet in neuer Betrachtungsweise.

3. Seminarausarbeitungen.

4. Vorträge von Tagungen.

Ferner kommen auch ältere vergriffene spezielle Vorlesungen, Seminare und Berichte in Frage, wenn nach ihnen eine anhaltende Nachfrage besteht.

Die Beiträge dürfen im Interesse einer größeren Aktualität durchaus den Charakter des Unfertigen und Vorläufigen haben. Sie brauchen Beweise unter Umständen nur zu skizzieren und dürfen auch Ergebnisse enthalten, die in ähnlicher Form schon erschienen sind oder später erscheinen sollen.

Die Herausgabe der „*Lecture Notes*" Serie durch den Springer-Verlag stellt eine Dienstleistung an die mathematischen Institute dar, indem der Springer-Verlag für ausreichende Lagerhaltung sorgt und einen großen internationalen Kreis von Interessenten erfassen kann. Durch Anzeigen in Fachzeitschriften, Aufnahme in Kataloge und durch Anmeldung zum Copyright sowie durch die Versendung von Besprechungsexemplaren wird eine lückenlose Dokumentation in den wissenschaftlichen Bibliotheken ermöglicht.

Lecture Notes in Mathematics
A collection of informal reports and seminars
Edited by A. Dold, Heidelberg and B. Eckmann, Zürich

Series: Mathematisches Institut der Universität Bonn · Adviser: F. Hirzebruch

52

D. J. Simms
Trinity College, Dublin

Lie Groups and Quantum Mechanics

1968

Springer-Verlag · Berlin · Heidelberg · New York

Library of Congress Catalog Card Number 68 - 24468 Title No. 7372

Preface

These notes are based on a series of twelve lectures given at the University of Bonn in the Wintersemester 1966-67 to a mixed group of mathematicians and theoretical physicists. I am grateful to Professors Hirzebruch, Bleuler and Klingenberg for giving me the opportunity of speaking at their seminar.

These notes are written primarily for the mathematician who has an elementary acquaintance with Lie groups and Lie algebras and who would like an account of the ideas which arise from the concept of relativistic invariance in quantum mechanics. They may also be of interest to the theoretical physicist who wants to see familiar material presented in a form which uses standard concepts from other areas of mathematics.

The presentation owes very much to the lecture notes of Mackey [24] and [25] and of Hermann [15] . Many of the original ideas are due to Wigner and Bargmann [37] , [1] and [39] . A useful collection of reprints is contained in Dyson [11] .

I should like to thank Professor F. Hirzebruch for his help and stimulation, Dr. D. Arlt for useful criticisms, and the Mathematisches Institut Bonn for support during this time and for the typing of the manuscript.

Dublin, April 1967 D. J. Simms

Contents

Section 1. Relativistic Invariance.

Causality.

Let M be the set of all space-time events. Any choice of observer defines a bijective map $M \to R^4$, $x \mapsto (x_1, x_2, x_3, x_4)$ where x_1, x_2, x_3 are space coordinates and x_4 is the time coordinate of the event x as seen by the chosen observer. This gives M the structure of a real 4-dimensional vector space with indefinite Lorentz scalar product

$$\langle x,y \rangle = - x_1y_1 - x_2y_2 - x_3y_3 + x_4y_4 ,$$

called the Minkowski structure of M relative to the given observer. The scalar product $\langle x,y \rangle$ may be given the following physical interpretation: $\sqrt{\langle x - y, x - y \rangle}$ is the time interval between events x and y as measured by a clock which moves with uniform velocity relative to our observer and is present at both events. Let us write $x < y$ if the event x is able to influence the event y, in the eyes of our observer. This means that y occurs later in time then x and that a physical body such as a clock is able to be present at both events. Thus we define:

$$x < y \text{ if and only if } x_4 < y_4 \text{ and } \langle x - y , x - y \rangle > 0 .$$

This partial order on M expresses the idea of causality, as seen by our chosen observer. An event x is time-like if $\langle x,x \rangle > 0$. The relation $\langle x,y \rangle > 0$ is an equivalence relation on the set of all time-like events, with two equivalence classes: the future and past events. Moreover $x < y$ if and only if $y - x$ is a future time-like event.

A change of observer determines a bijective map $f : M \longrightarrow M$, where $f(x)$ is the event which appears to the new observer to be the same as the event x does to the old observer.

The diagram

$$\begin{array}{ccc} M & \xrightarrow{\;f\;} & M \\ & {\scriptstyle \text{old}}\searrow \quad \swarrow{\scriptstyle \text{new}} & \\ & \mathbb{R}^4 & \end{array}$$

commutes. Event $f(x)$ will influence event $f(y)$, according to the new observer, if and only if x influences y according to the old observer. The two observers will have the same idea of causality provided that

$$x < y \Longleftrightarrow f(x) < f(y)$$

for all $x, y \in M$. In this case we call f a <u>causal automorphism</u> relative to our observer. If the idea of causality is to be preserved, we must limit ourselves to observers which are related to our chosen observer by a causal automorphism.

Let $\mathbb{R}^*$ denote the multiplicative group of non-zero real numbers. A <u>dilatation</u> of M is a map $M \longrightarrow M$ of the form $x \longmapsto \lambda x$, $\lambda \in \mathbb{R}^*$. A <u>translation</u> of M is a map $x \longmapsto x + a$, $a \in M$. A <u>homogeneous Lorentz transformation</u> of M is a linear map $A : M \longrightarrow M$ with $\langle Ax, Ay \rangle = \langle x, y \rangle$ for all $x, y \in M$. Here M is given its Minkowski structure defined by a chosen observer.

The group of translations may be identified with the additive group M. The group $\mathcal{L}$ of homogeneous Lorentz transformations admits two continuous homomorphisms $\xi, \eta : \mathcal{L} \longrightarrow \mathbb{R}^*$, where $\xi(A)$ is the determinant of A and $\eta(A) = \pm 1$ according to whether A leaves the equivalence classes of future and past events fixed or interchanges them. The <u>orthochronous homogeneous Lorentz group</u> $\mathcal{L}_t$ is the kernel of η. The elements of $\mathcal{L}_t$, as well as all dilatations and translations, are causal automorphisms. The <u>restricted Lorentz group</u> $\mathcal{L}_0$ is the intersection $\ker \xi \cap \ker \eta$, and is the connected component of the identity in $\mathcal{L}$.

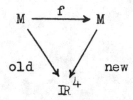

If a group G acts on a group H as a group of automorphisms, $h \mapsto h^g$, $g \in G$, then the cartesian product $H \times G$ with the group operation

$$(h_1,g_1)(h_2,g_2) = (h_1 h_2^{g_1}\ ,\ g_1 g_2)$$

is called a <u>semi-direct product</u> and denoted by $H \circledS G$. We note that $h \mapsto (h,1)$ and $g \mapsto (1,g)$ embed H as a normal subgroup and G as a subgroup of $H \circledS G$. Moreover

$$h^g = ghg^{-1}$$

in $H \circledS G$.

Any product of dilatations, translations, and orthochronous Lorentz transformations is of the form

$$x \mapsto A\ \lambda \cdot x + a$$

where $(a,A,\lambda) \in M \times \mathcal{L}_t \times \mathbb{R}^*$. The composition of $x \mapsto A\ \lambda \cdot x + a$ with $x \mapsto B\ \mu \cdot x + b$ is $x \mapsto AB\ \lambda \cdot \mu \cdot x + A\ \lambda \cdot b + a$, which corresponds to a group operation

$$(a,A,\lambda)\cdot(b,B,\mu) = (a + A\ \lambda \cdot b\ ,\ AB\ ,\ \lambda \cdot \mu\ ).$$

This shows that the group generated by such transformations is a semi-direct product

$$M \circledS (\mathcal{L}_t \times \mathbb{R}^*)$$

relative to the natural action of $\mathcal{L}_t \times \mathbb{R}^*$ on M.

Zeeman [40] has shown that $M \circledS (\mathcal{L}_t \times \mathbb{R}^*)$ is the complete group of causal automorphisms of M.

Causal invariance in quantum mechanics.

The pure states of a quantum mechanical system are represented by the projective space $\hat{H}$ of 1-dimensional subspaces of a separable complex Hilbert space H. For each non-zero vector ψ in H, let

$$\hat{\psi} = \{\lambda\psi \mid \lambda \in \mathbb{C}\}$$

be the 1-dimensional subspace containing ψ.

The inner product $\langle\psi,\varphi\rangle$ on the Hilbert space H defines a real valued function

$$\langle\hat{\psi},\hat{\varphi}\rangle = \frac{|\langle\psi,\varphi\rangle|^2}{\|\psi\|^2\|\varphi\|^2}$$

on $\hat{H} \times \hat{H}$. The physical interpretation is that $\langle\hat{\psi},\hat{\varphi}\rangle$ is the transition probability, the probability of finding the system to be in the state $\hat{\psi}$ when it is in the state $\hat{\varphi}$.

A bijective map $T: \hat{H} \to \hat{H}$ is an automorphism if it preserves the transition probability:

$$\langle T\hat{\psi}, T\hat{\varphi}\rangle = \langle\hat{\psi},\hat{\varphi}\rangle$$

for all $\hat{\psi}, \hat{\varphi} \in \hat{H}$.

If A is a unitary or anti-unitary transformation of H then it defines an automorphism $\hat{A}$ of $\hat{H}$ by:

$$\hat{A}\,\hat{\psi} = \widehat{(A\,\psi)}\ .$$

By an <u>anti-unitary</u> transformation A we mean one such that

$$A(\psi + \varphi) = A\psi + A\varphi$$

$$A\lambda\psi = \bar{\lambda}A\psi$$

$$\langle A\psi, A\varphi\rangle = \overline{\langle \psi,\varphi\rangle}$$

for all $\psi, \varphi \in H$, $\lambda \in \mathbb{C}$. The product of two anti-unitary transformations is unitary, so that the group $U(H)$ of unitary transformations is a subgroup of index two of the group $\tilde{U}(H)$ of unitary or anti-unitary transformations. The map $e^{i\theta} \longmapsto e^{i\theta}\cdot 1$ embeds $U(1)$ as a subgroup of $U(H)$.

Let $\mathrm{Aut}(\hat{H})$ be the group of automorphisms of $\hat{H}$, and let $\pi:\tilde{U}(H) \to \mathrm{Aut}(\hat{H})$ be the map $\pi(A) = \hat{A}$. We have the fundamental result:

<u>Theorem of Wigner.</u> <u>The sequence</u>

$$1 \to U(1) \to \tilde{U}(H) \xrightarrow{\pi} \mathrm{Aut}(\hat{H}) \to 1 \qquad (1)$$

<u>is exact.</u>

This means that every automorphism of $\hat{H}$ is of the form $\hat{A}$ where A is a unitary or anti-unitary transformation of H. Moreover, if $\hat{A} = \hat{B}$ then $A = e^{i\theta}B$ with $e^{i\theta} \in U(1)$. For Wigner's proof see [38], appendix to chapter 2o. See also [2].

If $f: M \to M$ is a bijective map arising from a change of observer, it determines a bijective map $T_f: \hat{H} \to \hat{H}$, where $T_f \hat{\psi}$ is the quantum mechanical state which appears to the new observer to be the same as the state $\hat{\psi}$ does to the old observer. If $g: M \to M$ is another change of observer we shall suppose that

$$T_{fg} = T_f T_g \, .$$

A physical justification of this assumption could be based upon the relation between states and assemblies of events in space-time, and on the definition of f and g in terms of events in space-time.

The new observer will have the same idea of transition probability as the old observer if and only if T_f is an automorphism of $\hat{H}$. The group of all transformations of M which are associated with observers which have the same idea of causality as our chosen observer is $M \circledS (\mathcal{L}_t \times \mathbb{R}^*)$ by the result of Zeeman. If all the observers also have the same idea of transition probability, then we have a homomorphism

$$T: M \circledS (\mathcal{L}_t \times \mathbb{R}^*) \longrightarrow \mathrm{Aut}(\hat{H}) \, .$$

In this case we say that we have <u>causal invariance.</u> If the weaker condition holds that all observers which are related to our chosen observer by transformations in the <u>restricted inhomogeneous Lorentz group</u> $M \circledS \mathcal{L}_0$ have the same idea of transition probability then we have a homomorphism

$$T: M \circledS \mathcal{L}_0 \longrightarrow \mathrm{Aut}(\hat{H}) .$$

In this case we have relativistic invariance. From now on we will confine ourselves to relativistic invariance.

The exact sequence (1) gives an exact sequence

$$1 \longrightarrow U(1) \longrightarrow U(H) \xrightarrow{\pi} U(\hat{H}) \longrightarrow 1$$

where $U(\hat{H})$ is the image of $U(H)$ under π and is a subgroup of index two in $\mathrm{Aut}(\hat{H})$.

If G is a connected Lie group, then the image of any homomorphism

$$T: G \longrightarrow \mathrm{Aut}(\hat{H})$$

is contained in $U(\hat{H})$. To see this we note that the exponential map in G shows that there is a neighbourhood V of the identity in G in which each element is a square and is therefore mapped by T into $U(\hat{H})$.

Since $M \circledS \mathcal{L}_0$ is a connected Lie group, we have a homomorphism

$$T: M \circledS \mathcal{L}_0 \longrightarrow U(\hat{H})$$

in the case of relativistic invariance.

Continuity.

Let the Hilbert space H be given its usual norm topology, and let the projective space $\hat{H}$ be given the quotient topology relative to the surjection $\psi \rightarrow \hat{\psi}$. We assume that, for each fixed state $\hat{\psi}$, the state $T_f \hat{\psi}$ will depend continuously on the change of observer. This means that the map

$$M \circledS \mathcal{L}_0 \longrightarrow \hat{H}$$

$f \mapsto T_f \hat{\psi}$ is continuous for each $\hat{\psi} \in \hat{H}$. This is equivalent to the continuity of

$$T: M \circledS \mathcal{L}_0 \longrightarrow U(\hat{H})$$

where $U(\hat{H})$ is given the weakest topology such that all maps $U(\hat{H}) \longrightarrow \hat{H}$, $\hat{A} \longmapsto \hat{A}\hat{\psi}$, are continuous, $\hat{\psi} \in \hat{H}$. The same topology on $U(\hat{H})$ may be obtained by giving $U(H)$ the strong operator topology, which is the weakest topology such that all maps $U(H) \longrightarrow H$, $A \longmapsto A\psi$, are continuous, $\psi \in H$, and then taking the quotient topology relative to the surjection $A \longmapsto \hat{A}$. The equality of these two topologies on $U(\hat{H})$ follows from [1] theorem 1.1.

By a <u>projective representation</u> T of a topological group G we shall mean a continuous homomorphism $T: G \longrightarrow U(\hat{H})$ with the topology given above. By a <u>representation</u> T of G we mean a continuous homomorphism $T: G \longrightarrow U(H)$ where $U(H)$ is given the strong operator topology defined above.

Section 2. Lifting Projective Representations.

If τ is a representation of a topological group G in $U(H)$ then $\pi\circ\tau$ is a projective representation of G :

$$G \xrightarrow{\tau} U(H) \xrightarrow{\pi} U(\hat{H}) .$$

Conversely, if $T: G \longrightarrow U(\hat{H})$ is a projective representation then T <u>admits a lifting, τ</u> if there exists a representation $\tau: G \longrightarrow U(H)$ such that $T = \pi\circ\tau$.

Let G be a connected Lie group with simply connected covering group $\tilde{G}$ and covering map p with kernel K . Let T be a projective representation of G in $U(\hat{H})$. Using the theorem of Wigner we have the diagram

$$\begin{array}{ccccccccc} 1 & \longrightarrow & K & \longrightarrow & \tilde{G} & \xrightarrow{p} & G & \longrightarrow & 1 \\ & & & & & & \downarrow T & & \\ 1 & \longrightarrow & U(1) & \longrightarrow & U(H) & \xrightarrow{\pi} & U(\hat{H}) & \longrightarrow & 1 \end{array}$$

with both rows exact. $T\circ p$ is a projective representation of $\tilde{G}$ and $T\circ p\ (K) = 1$. Suppose $T\circ p$ admits a lifting $\sigma: \tilde{G} \longrightarrow U(H)$; then $\sigma(K) \subset U(1)$. Conversely any representation σ of $\tilde{G}$ in $U(H)$ such that $\sigma(K) \subset U(1)$ will define a unique projective representation T of G in $U(\hat{H})$ such that $T\circ p = \pi\circ\sigma$.

These considerations show that, if the simply connected Lie group $\tilde{G}$ has the property that every projective representation of $\tilde{G}$ admits a lifting, then the determination of all the projective representations of G is equivalent to the determination of the representations of $\tilde{G}$

which map K into U(1) . Bargmann in [1] Theorem 3.2, Lemma 4.9, and § 2, proved that the possibility of lifting projective representations of $\widetilde{G}$ depends on the cohomology of the Lie algebra of $\widetilde{G}$. The remainder of this section will be devoted to giving a topological proof of Bargmann's result. We shall prove four theorems, the fourth being the theorem of Bargmann.

Theorem 1. $U(H)$ is a principal bundle with base space $U(\hat{H})$, projection π and fibre $U(1)$.

Proof. For the relevant definitions we refer to [17] I. 3.2.a) and also [19] III. 4. We first note that although $U(H)$ is not a topological group in general since the multiplication $U(H) \times U(H) \longrightarrow U(H)$ is not continuous, see [28] § 33.2, the multiplication $U(1) \times U(H) \longrightarrow U(H)$ is continuous. It follows that π is an open map.

For each non-zero $\varphi \in H$ the set

$$V_\varphi = \{A \mid < A\varphi,\varphi > \neq 0\}$$

is open in $U(H)$, and such sets give an open cover of $U(H)$. The sets $W_\varphi = \pi(V_\varphi)$ therefore form an open cover of $U(\hat{H})$, and $\pi^{-1}(W_\varphi) = V_\varphi$.

Let $\eta_\varphi : V_\varphi \longrightarrow U(1)$ be defined by

$$\eta_\varphi(A) = \frac{< A\varphi,\varphi >}{|< A\varphi,\varphi >|} .$$

Then $\eta_\varphi(\lambda A) = \lambda\cdot\eta_\varphi(A)$ for each $\lambda \in U(1)$. Let $h_\varphi : V_\varphi \longrightarrow W_\varphi \times U(1)$ be defined by

$$h_\varphi(A) = (\pi A\ ,\ \eta_\varphi(A)) .$$

This is continuous with continuous inverse

$$(\pi A\ ,\ e^{i\theta}) \longmapsto e^{i\theta}[\eta_\varphi(A)]^{-1}A$$

and is therefore a homeomorphism. This proves the bundle space property that $\pi^{-1}(W_\varphi)$ is homeomorphic to the topological product $W_\varphi \times U(1)$.

It remains to show that the structure group is $U(1)$. If ψ, φ are non-zero vectors in H and if $\pi A \in W_\psi \cap W_\varphi$ and $e^{i\theta} \in U(1)$ then

$$\begin{aligned} h_\psi h_\varphi^{-1}\ (\pi A, e^{i\theta}) &= h_\psi\ \{\ e^{i\theta}\eta_\varphi(A)^{-1}A\ \} \\ &= (\pi A\ ,\ e^{i\theta}\eta_\varphi(A)^{-1}\eta_\psi(A))\ . \end{aligned}$$

Thus the fibre coordinate $e^{i\theta}$ is multiplied by $\eta_\varphi(A)^{-1}\eta_\psi(A) \in U(1)$ on change of coordinate neighbourhood.
Moreover the function

$$W_\psi \cap W_\varphi \longrightarrow U(1)\ ,$$

$\pi A \longmapsto \eta_\varphi(A)^{-1}\eta_\psi(A)$ is continuous. This completes the proof.

<u>Theorem 2.</u> <u>Any continuous map</u> $T\colon G \longrightarrow U(\hat{H})$ <u>of a connected simply connected Lie group</u> G <u>can be lifted to a continuous map</u> $\tau\colon G \longrightarrow U(H)$ <u>with</u> $T = \pi\circ\tau$.

<u>Proof.</u> T induces a $U(1)$ bundle, $E \xrightarrow{\sigma} G$, with base space G , total space

$$E = \{(g,A) \mid T(g) = \pi(A)\ \} \subset G \times U(H)$$

and projection $\sigma(g,A) = g$. See [17] I. 3.3. or [33] page 98 .
The diagram

$$\begin{array}{ccc} E & \xrightarrow{\alpha} & U(H) \\ {\scriptstyle\sigma}\downarrow & & \downarrow{\scriptstyle\pi} \\ G & \xrightarrow[T]{} & U(\hat{H}) \end{array}$$

commutes, where $\alpha(g,A) = A$. By a generalisation [3] 5., 17., and 18., of a result of Cartan, the second homotopy group of G is zero. Since G is also simply connected, it follows from a theorem of Hurewicz [19] II . Corollary 9.2., and by the universal coefficient theorem [12] V. Exercise G. 3., that the singular cohomology group $H^2(G,\mathbb{Z})$ is zero. Since the first homotopy group of the fibre $U(1)$ is $\mathbb{Z}$, it follows from obstruction theory [35] 35.5 and 29.8 that a continuous map $s: G \longrightarrow E$ exists with $\sigma \circ s = 1$.

The continuous map $\tau = \alpha \circ s$ has the property required since $\pi \circ \tau = \pi \circ \alpha \circ s = T \circ \sigma \circ s = T$. This completes the proof.

<u>Extensions and factor sets.</u>

Let G and K be Lie groups with K abelian. A <u>factor set</u> for the pair (G,K) is a continuous map

$$\omega: G \times G \longrightarrow K$$

such that $\omega(1,1) = 1$, and

$$\omega(x,y)\cdot\omega(xy,z) = \omega(x,yz)\cdot\omega(y,z)$$

for all $x, y, z \in G$.

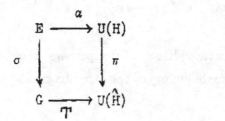

Let E^ω be the topological group which as a topological space is the product $G \times K$ and has group operation

$$(x_1,k_1)(x_2,k_2) = (x_1x_2,\ \omega(x_1,x_2)k_1k_2) .$$

The properties of ω ensure that E^ω is a topological group, with K imbedded in the centre of E^ω by the map $k \longmapsto (1,k)$. Since E^ω is locally Euclidean it is a Lie group by the well known result [27] 2.15 of Montgomery, Zippin and Gleason, and we have an exact sequence of Lie groups

$$1 \longrightarrow K \longrightarrow E^\omega \xrightarrow{\ \alpha\ } G \longrightarrow 1$$

where $\alpha(x,k) = x$.

Let LG and LK be the Lie algebras of G and K . A skew-symmetric bilinear map θ: LG × LG ⟶ LK is called a <u>factor set</u> for the pair (LG,LK) if

$$\theta(x,[y,z]) + \theta(y,[z,x]) + \theta(z,[x,y]) = 0$$

for all $x, y, z \in LG$. The factor set θ is <u>trivial</u> if there exists a linear map ϑ: LG ⟶ LK with

$$\theta(x,y) = \vartheta(\,[x,y]\,)$$

for all $x, y \in LG$. The quotient of the additive group of factor sets by the subgroup of trivial factor sets is denoted by $H^2(LG,LK)$ and called the <u>2nd cohomology group</u> of LG with coefficients in the vector space LK (with trivial action of LG on LK) . See [20] III. 10. and [6] XIII. 8. for more details.

Consider now the exact sequence of Lie algebras

$$0 \longrightarrow LK \longrightarrow LE^{\omega} \xrightarrow{\dot{\alpha}} LG \longrightarrow 0 \ .$$

Choose a linear map $\beta\colon LG \longrightarrow LE^{\omega}$ such that $\dot{\alpha}\circ\beta = 1$. Put

$$\theta(x,y) = [\beta(x) \ , \beta(y)] - \beta([x,y])$$

for each $x, y \in LG$. This is a factor set for the pair (LG,LK) . If θ is a trivial factor set then $\theta(x,y) = \vartheta([x,y])$ where $\vartheta\colon LG \longrightarrow LK$ is linear, so that

$$\mu(x) = \beta(x) + \vartheta(x)$$

defines a homomorphism $\mu\colon LG \longrightarrow LE^{\omega}$ with $\dot{\alpha}\circ\mu = 1$. If G is connected and simply connected, then there is a homomorphism $\gamma\colon G \longrightarrow E^{\omega}$ such that $\dot{\gamma} = \mu$ and $\alpha\circ\gamma = 1$. Thus γ must be of the form

$$\gamma(x) = (x,\lambda(x))$$

where λ is a continuous map $G \longrightarrow K$. Since γ is a homomorphism,

$$\begin{aligned}(xy,\lambda(xy)) &= (x,\lambda(x))(y,\lambda(y)) \\ &= (xy,\omega(x,y)\lambda(x)\lambda(y))\end{aligned}$$

so that

$$\lambda(xy) = \omega(x,y)\lambda(x)\lambda(y)$$

for all $x, y \in G$.

We have now proved the following;

Theorem 3. *Let* G *and* K *be Lie groups,* G *connected and simply connected, and* K *abelian. Let* $H^2(LG, LK) = 0$. *Then for any factor set* ω *for* (G, K) *there is a continuous map* $\lambda : G \longrightarrow K$ *with*

$$\lambda(xy) = \omega(x,y)\lambda(x)\lambda(y)$$

for all $x, y \in G$.

We are now ready to prove the main result on lifting projective representations.

Theorem 4 (Bargmann). *Let* G *be a connected and simply connected Lie group with* $H^2(LG, \mathbb{R}) = 0$. *Then any projective representation* $T: G \longrightarrow U(\hat{H})$ *admits a lifting* $\tau: G \longrightarrow U(H)$ *which is a representation.*

Proof. By theorem 2 there exists a continuous map $\sigma: G \longrightarrow U(H)$ with $\pi \circ \sigma = T$.

$$\begin{array}{ccccccccc} & & & & G & & & & \\ & & & \sigma \swarrow & & \searrow T & & & \\ 1 & \longrightarrow & U(1) & \longrightarrow & U(H) & \xrightarrow[\pi]{} & U(\hat{H}) & \longrightarrow & 1 \end{array}$$

We can (and will) choose σ so that $\sigma(1) = 1$.

For each $x, y \in G$ we have

$$\pi(\sigma(x)\sigma(y)) = \pi\circ\sigma(x).\pi\circ\sigma(y) = T(x).T(y) = T(xy) = \pi\circ\sigma(xy) .$$

Therefore

$$\sigma(x)\sigma(y) = \omega(x,y)\sigma(xy)$$

where $\omega(x,y) \in U(1)$. The map $\omega: G \times G \longrightarrow U(1)$ is continuous. Indeed for any unit vector ψ in H we have

$$[\omega(x',y') - \omega(x,y)]\sigma(x'y')\psi$$

$$= \omega(x,y)[\sigma(xy) - \sigma(x'y')]\psi + \sigma(x')[\sigma(y') - \sigma(y)]\psi + [\sigma(x') - \sigma(x)]\sigma(y)\psi$$

so that taking norms

$$|\omega(x',y') - \omega(x,y)| \leq \| \sigma(xy)\psi - \sigma(x'y')\psi\|$$

$$+\| \sigma(y')\psi - \sigma(y)\psi \|$$

$$+\| \sigma(x')\psi - \sigma(x)\psi \| .$$

The continuity of ω follows from the continuity of the maps $(x,y)\longmapsto\sigma(xy)\psi$, $\sigma(y)\psi$, $\sigma(x)\psi$. Moreover ω satiesfies the conditions for a factor set for the pair $(G,U(1))$. Since the Lie algebra of $U(1)$ is $\mathbb{R}$ we can apply theorem 3 to obtain a continuous map $\lambda: G\longrightarrow U(1)$ with

$$\lambda(xy) = \omega(x,y)\lambda(x)\lambda(y) .$$

Now put $\tau(x) = \lambda(x)\cdot\sigma(x)$. τ is the required representation lifting T .

Section 3. The Relativistic Free Particle.

We have seen in section 1 that the relativistic invariance of a quantum mechanical system requires a projective representation

$$T: M \circledS \mathcal{L}_0 \longrightarrow U(\hat{H})$$

of the restricted inhomogeneous Lorentz group. A permissible choice of observer defines a bijective map $M \xrightarrow{\alpha} \mathbb{R}^4$ which is a vector space isomorphism preserving the Lorentz scalar products. The group $O(3,1)$ of homogeneous Lorentz transformations of $\mathbb{R}^4$ relative to its scalar product

$$\langle x,y\rangle = -x_1y_1 - x_2y_2 - x_3y_3 + x_4y_4 = x'\Lambda x$$

is the group of 4×4 real matrices A with $\langle Ax,Ay\rangle = \langle x,y\rangle$. Thus

$$O(3,1) = \{ A \mid A' \cdot \Lambda \cdot A = \Lambda \}$$

where Λ is the diagonal matrix with entries $-1, -1, -1, 1$. The group of translations of $\mathbb{R}^4$ is naturally isomorphic to $\mathbb{R}^4$, and the group of inhomogeneous Lorentz transformations of $\mathbb{R}^4$ is the semi-direct product

$$\mathbb{R}^4 \circledS O(3,1)$$

relative to the natural action of $O(3,1)$ on $\mathbb{R}^4$.

An inhomogeneous Lorentz transformation of M, $x \longmapsto Ax + a$ will appear to the observer as a transformation of $\mathbb{R}^4$, $\alpha(x) \longmapsto \alpha(Ax) + \alpha(a) = (\alpha A \alpha^{-1})\alpha(x) + \alpha(a)$. The map

$$(a,A) \longmapsto (\alpha(a), \alpha A \alpha^{-1})$$

is an isomorphism α_* of $M \circledS \mathcal{L}$ onto $\mathbb{R}^4 \circledS O(3,1)$. Let $SO(3,1)$ be the connected component of the identity in $O(3,1)$, then $M \circledS \mathcal{L}_0$ is mapped isomorphically onto $\mathbb{R}^4 \circledS SO(3,1)$.

The simply connected covering group of $SO(3,1)$ is $SL(2,\mathbb{C})$, the group of 2×2 complex matrices with determinant 1. The covering map

$$SL(2,\mathbb{C}) \xrightarrow{\ \eta\ } SO(3,1)$$

can be defined as follows. Let

$$\tau_1 = \begin{pmatrix} 0 & 1 \\ 1 & 0 \end{pmatrix}, \quad \tau_2 = \begin{pmatrix} 0 & -i \\ i & 0 \end{pmatrix}, \quad \tau_3 = \begin{pmatrix} 1 & 0 \\ 0 & -1 \end{pmatrix}, \quad \tau_4 = \begin{pmatrix} 1 & 0 \\ 0 & 1 \end{pmatrix}$$

Identify each $x = (x_1, x_2, x_3, x_4) \in \mathbb{R}^4$ with the Hermitian matrix

$$\underset{\sim}{x} = x_1\tau_1 + x_2\tau_2 + x_3\tau_3 + x_4\tau_4 = \begin{pmatrix} x_4 + x_3 & x_1 - ix_2 \\ x_1 + ix_2 & x_4 - x_3 \end{pmatrix},$$

so that det $\underset{\sim}{x} = \langle x,x \rangle$. For each $A \in SL(2,\mathbb{C})$ let

$$\eta(A) : \underset{\sim}{x} \longrightarrow A\underset{\sim}{x}A^* ,$$

where A^* is the inverse transpose of A . Since

$$\det(A\underset{\sim}{x}A^*) = \det \underset{\sim}{x}$$

it follows that $\eta(A) \in O(3,1)$. Since η is continuous it maps the connected group $SL(2,\mathbb{C})$ into the connected group $SO(3,1)$. Finally, η is surjective [13] Part II section 1 § 5 , with discrete kernel $\{1,-1\}$.

The simply connected cover of $\mathbb{R}^4 \circledS SO(3,1)$ is therefore $\mathbb{R}^4 \circledS SL(2,\mathbb{C})$, relative to the action $\underset{\sim}{x} \longrightarrow A\underset{\sim}{x}A^*$ of $SL(2,\mathbb{C})$ on $\mathbb{R}^4$.

Let $\tilde{\mathcal{L}}_0$ be the simply connected covering group of $\mathcal{L}_0$, so that $M \circledS \tilde{\mathcal{L}}_0$ is the simply connected cover of $M \circledS \mathcal{L}_0$. The isomorphism $\alpha_*: M \circledS \mathcal{L}_0 \longrightarrow \mathbb{R}^4 \circledS SO(3,1)$ induces an isomorphism $\alpha_*: M \circledS \tilde{\mathcal{L}}_0 \longrightarrow \mathbb{R}^4 \circledS SL(2,C)$.

Bargmann [1] (6.17) has shown that the cohomology group $H^2(LG,\mathbb{R}) = 0$ where LG is any one of the isomorphic Lie algebras of $M \circledS \mathcal{L}_0$, $M \circledS \tilde{\mathcal{L}}_0$, $\mathbb{R}^4 \circledS SO(3,1)$, $\mathbb{R}^4 \circledS SL(2,C)$. By the results of section 2 we can conclude that the projective representation T is induced by a representation

$$\tau: M \circledS \tilde{\mathcal{L}}_0 \longrightarrow U(H) .$$

such that $\tau(\{1,-1\}) \subset U(1)$. If τ is irreducible then we call the quantum mechanical system an <u>elementary relativistic free particle.</u>

Section 4. Lie Algebras and Physical Observables.

Let $T : G \longrightarrow U(H)$ be a representation of a Lie group G in a Hilbert space H, and let LG be the Lie algebra of G. For each $X \in LG$ the 1-parameter subgroup $t \longmapsto \exp tX$ is mapped into a (strongly continuous) 1-parameter group $t \longmapsto T\circ\exp tX$ of unitary operators on H. By the fundamental theorem of Stone [34] Theorem B there is a unique skew-adjoint operator $\dot{T}(X)$ on H such that

$$T\circ \exp tX = \exp t\dot{T}(X)$$

for all $t \in \mathbb{R}$. For $\psi \in H$, $\dot{T}(X)\psi$ is the derivative at $t = 0$ of the map

$$\mathbb{R} \longrightarrow H , \ t \longmapsto (T\circ\exp tX)\psi .$$

The domain of the operator $\dot{T}(X)$ is the set of ψ for which this derivative exists.

There exists a dense set D_T in H on which $\dot{T}(X)$ is defined and essentially skew-adjoint for all $X \in LG$; each $\dot{T}(X)$ is therefore determined by its restriction to D_T. Let $S(D_T)$ be the Lie algebra of skew-symmetric operators having D_T as common invariant domain. Then

$$\dot{T}: LG \longrightarrow S(D_T)$$

is a Lie algebra homomorphism. This implies for instance that if $X,Y \in LG$ then $\dot{T}(X)\dot{T}(Y) - \dot{T}(Y)\dot{T}(X)$ is essentially self adjoint on D_T with unique self adjoint extension $\dot{T}[X,Y]$. For these results see for instance [31], Theorem 3.1.

Suppose now that the associated projective space $\hat{H}$ is the system of pure states of a quantum mechanical system. Each self adjoint operator A on H is then associated with a physical observable. In fact the spectral theorem for A gives a projection valued measure P on $\mathbb{R}$, and

$$\langle P(E)\psi,\psi\rangle$$

is the probability of the value of that observable being found to lie in the Borel set $E \subset \mathbb{R}$ when the system is in the state $\hat{\psi}$ determined by the unit vector ψ. For an account see [26], 2-2.

The self adjoint operator $\frac{1}{i}\cdot\dot{T}(X)$ must therefore represent a physical quantity. We say we have a <u>physical interpretation</u> of the symmetry group G when we have specified which physical quantity is associated with X, for each $X \in LG$.

<u>Physical interpretation of the inhomogeneous Lorentz group</u>.

The inhomogeneous Lorentz group $M \circledS \mathcal{L}$ is a semi-direct product. Its Lie algebra is therefore a <u>semi-direct sum</u>

$$LM \circledS L\mathcal{L}$$

where the Lie algebra LM of the vector group M is abelian and naturally identifiable with M itself, and the Lie algebra $L\mathcal{L}$ is the algebra of linear transformations of M which are skew relative to the Lorentz scalar product.

$$L\mathcal{L} = \{ A \mid \langle Ax,y\rangle + \langle x,Ay\rangle = 0 \text{ , all } x, y \in M \} .$$

The semi-direct sum is relative to the natural action of $L\mathscr{L}$ on $LM \approx M$. This means that $LM \circledS L\mathscr{L} = LM \oplus L\mathscr{L}$ as a vector space, with Lie product

$$[(a,A),(b,B)] = ([a,b] + Ab - Ba, [A,B]) = (Ab - Ba, [A,B]) .$$

See [18] for material on semi-direct sums.

A choice of observer $M \xrightarrow{\alpha} \mathbb{R}^4$ gives a Lie group isomorphism

$$\alpha_*: M \circledS \mathscr{L} \longrightarrow \mathbb{R}^4 \circledS O(3,1)$$

which induces a Lie algebra isomorphism

$$\alpha_*: LM \circledS L\mathscr{L} \longrightarrow L\mathbb{R}^4 \circledS LO(3,1) .$$

$L\mathbb{R}^4$ is abelian and identifiable with $\mathbb{R}^4$ as a vector space. $LO(3,1)$ is the Lie algebra of 4×4 real matrices which are skew relative to the Lorentz scalar product on $\mathbb{R}^4$:

$$LO(3,1) = \{ A \mid A'\Lambda + \Lambda A = 0 \} .$$

$L\mathbb{R}^4 \circledS LO(3,1)$ is the semi-direct sum relative to the natural action of $LO(3,1)$ on $L\mathbb{R}^4 \approx \mathbb{R}^4$.

Rotations of $\mathbb{R}^4$ in the x_1x_2-plane give a 1-parameter subgroup of $O(3,1)$:

$$t \mapsto \begin{pmatrix} \cos t & \sin t & & \\ -\sin t & \cos t & & \\ & & 1 & \\ & & & 1 \end{pmatrix} = \exp t\cdot \begin{pmatrix} 0 & 1 & & \\ -1 & 0 & & \\ & & 0 & \\ & & & 0 \end{pmatrix} = \exp tm_{12}$$

with $m_{12} \in LO(3,1)$. Pure Lorentz rotations in the x_3x_4-plane give a 1-parameter subgroup of $O(3,1)$:

$$t \mapsto \begin{pmatrix} 1 & & & \\ & 1 & & \\ & & \cosh t & \sinh t \\ & & \sinh t & \cosh t \end{pmatrix} = \exp t\cdot \begin{pmatrix} 0 & & & \\ & 0 & & \\ & & 0 & 1 \\ & & 1 & 0 \end{pmatrix} = \exp tm_{14}$$

with $m_{14} \in LO(3,1)$. Translations of $\mathbb{R}^4$ in the x_1 direction give a 1-parameter subgroup :

$$t \longmapsto (t,0,0,0) = \exp t\pi_1$$

with $\pi_1 \in L\mathbb{R}^4$. In this way a basis

$$m_{23}, m_{31}, m_{12}, m_{14}, m_{24}, m_{34}$$

for $LO(3,1)$ is defined, and a basis

$$\pi_1, \pi_2, \pi_3, \pi_4$$

for $\mathbb{R}^4$; the latter corresponds to the standard basis for $\mathbb{R}^4$ under the isomorphism $L\mathbb{R}^4 \approx \mathbb{R}^4$. We associate π_1, π_2, π_3 with the components of <u>linear momentum</u> in the x_1, x_2, x_3 directions respectively, π_4 with <u>energy</u>, m_{23}, m_{31}, m_{12} with the components of <u>angular momentum</u> about the x_1, x_2, x_3 axes respectively, and m_{14}, m_{24}, m_{34} with components of <u>relativistic angular momentum</u>. We are now ready to give a physical interpretation of the simply connected covering group $G = M \circledS \tilde{\mathcal{L}}_0$ of the restricted inhomogeneous Lorentz group, as follows. Let $T: G \longrightarrow U(H)$ be the representation defined by relativistiv invariance. Let $\alpha_*: LG \longrightarrow L\mathbb{R}^4 \circledS LO(3,1)$ be the isomorphism defined by choice of observer $M \xrightarrow{\alpha} \mathbb{R}^4$. If $X \in LG$ and $\alpha_*(X) = \pi_1$ then we interpret $\frac{1}{i}\cdot\dot{T}(X)$ as the self-adjoint operator corresponding to linear momentum in the x_1 direction as seen by this observer. Similarly for the other components of linear momentum, the energy, and the angular momentum. LG is the Lie algebra of <u>relativistic observables.</u>

<u>Effect of change of observer</u>.

Let $f: M \longrightarrow M$ be a change of observer. The transformation $fgf^{-1}: M \longrightarrow M$ appears the same to the new observer as the transformation $g: M \longrightarrow M$ did to the old observer. If we limit ourselves to relativistic changes of observer then f is an inhomogeneous Lorentz transformation and $g \longmapsto fgf^{-1}$ is an inner automorphism $I(f)$ of $M \circledS \mathcal{L}$. The map

$$I: M \circledS \mathcal{L} \longrightarrow \operatorname{Aut}(M \circledS \mathcal{L})$$

is a homomorphism into the group of automorphisms of $M \circledS \mathcal{L}$.

The physical interpretation of this is that $I(f)$ is the transformation of $M \circledS \mathcal{L}$ induced by a change f of observer.

We note that if $A \in \mathcal{L}$ and $(b,B) \in M \circledS \mathcal{L}$ then $I(A)(b,B) = (0,A)(b,B)(0,A)^{-1} = (Ab,ABA^{-1})$. This shows that the action of $I(\mathcal{L})$ on M is the natural action of $\mathcal{L}$ on M, and that the action of $I(\mathcal{L})$ on $\mathcal{L}$ is induced by the natural action of $\mathcal{L}$ on $\mathrm{Hom}(M,M) \supset \mathcal{L}$.

The automorphism $I(f)$ of $M \circledS \mathcal{L}$ induces a Lie algebra automorphism

$$\mathrm{Ad}_f : LM \circledS L\mathcal{L} \longrightarrow LM \circledS L\mathcal{L}$$

where Ad_f is the derivative of $I(f)$ at the identity. The physical interpretation of this is that $\mathrm{Ad}_f X$ is the physical quantity which appears the same to the new observer as X did to the old observer. Thus the self-adjoint operator $\frac{1}{i}\cdot\dot{T}(\mathrm{Ad}_f X)$ represents the same quantity to the new observer as $\frac{1}{i}\cdot\dot{T}(X)$ does to the old.

We thus have a Lie group homomorphism, the <u>adjoint representation</u>

$$\mathrm{Ad}: G \longrightarrow \mathrm{Aut}\ (LG)$$

of the Lie group $G = M \circledS \mathcal{L}$. This expresses the action of the group G of changes of observer on the Lie algebra LG of relativistic observables.

The action of $\mathrm{Ad}(\mathcal{L})$ on LM is the natural action of $\mathcal{L}$ on $LM \approx M$. This action preserves the Lorentz scalar product on $LM \approx M$ and the elements of LM are therefore first order Lorentz tensors. The action of $\mathrm{Ad}(\mathcal{L})$ on $L\mathcal{L}$ is induced by the natural action of $\mathcal{L}$ on $\mathrm{Hom}(M,M) \supset L\mathcal{L}$. The elements of $L\mathcal{L}$ are therefore 2nd order skew Lorentz tensors.

Section 5. Universal Enveloping Algebra.

In the last section we have seen how the group $G = M \circledS \mathcal{L}_0$ of relativistic changes of observer acts on the Lie group LG of relativistic observables. It is of interst to know how G acts on polynomials in the non-commuting operators $\dot{T}(X)$, and in particular to find operators which are independent of a choice of observer. The natural setting for this is the universal enveloping algebra ULG which is an associative algebra containing LG as a vector subspace in such a way that the Lie product $[X,Y]$ equals the commutator $XY - YX$.

Tensor algebra

Let L be a real vector space and $TL = \mathbb{R} \oplus L \oplus (L \otimes L) \oplus (L \otimes L \otimes L) \oplus \ldots$ be the tensor algebra over L . TL may be characterised by the universal property:

i) TL is an associative algebra (with identity) containing L as a vector subspace, and generated by L .

ii) any linear map of L into an associative algebra A extends to a (unique) homomorphism of TL into A .

In particular any linear map $\lambda: L \to L$ extends to a unique automorphism $\lambda_a: TL \longrightarrow TL$. We shall need the fact that λ also extends to a unique derivation $\lambda_d: TL \longrightarrow TL$.

By a <u>derivation</u> D of a (not necessarily associative) algebra A we mean a linear map $D: A \to A$ with

$$D(xy) = (Dx)y + x(Dy)$$

all $x, y \in A$.

λ_a is characterised by $\lambda_a(x_1 \otimes .. \otimes x_r) = \lambda(x_1) \otimes ... \otimes \lambda(x_r)$, and λ_d by $\lambda_d(x_1 \otimes .. \otimes x_r) = \lambda(x_1) \otimes x_2 \otimes .. \otimes x_r + ... + x_1 \otimes .. .. \otimes x_{r-1} \otimes \lambda(x_r)$. For these facts see [4], [8] and [1o].

Universal enveloping algebra.

Let L be a real Lie algebra. The <u>universal enveloping algebra</u> UL is the quotient of the tensor algebra TL by the ideal J generated by the elements

$$X \otimes Y - Y \otimes X - [X,Y]$$

$X, Y \in L$. UL may be characterised by the universal property:

i) UL is an associative algebra (with identity) containing L as a vector subspace, generated by L, and with

$$[X,Y] = XY - YX$$

each $X, Y \in L$.

ii) any linear map λ of L into an associative algebra A with $\lambda[X,Y] = \lambda(X)\lambda(Y) - \lambda(Y)\lambda(X)$ extends to a (unique) homomorphism $UL \longrightarrow A$.

In effect UL is the algebra of all non-commutative polynomials in the elements of L , subject to the condition that the Lie product $[X,Y]$ in L equals the commutator $XY - YX$ in UL . Any automorphism $\lambda\colon L \longrightarrow L$ induces a (unique) automorphism of TL leaving the ideal J invariant and hence induces a (unique) automorphism $\lambda_a\colon UL \longrightarrow UL$. Any derivation $\lambda\colon L \longrightarrow L$ induces a derivation of TL leaving J invariant and thus a unique derivation $\lambda_d\colon UL \longrightarrow UL$.

For more details see [20] V. and [30] LA 3 .

<u>Symmetric algebra.</u>

Let L be a real vector space. The <u>symmetric algebra</u> SL of L is the quotient of the tensor algebra TL by the ideal I generated by the elements of the form

$$x \otimes y - y \otimes x$$

$x, y \in L$. SL may be characterised by the universal property:

i) SL is an associative commutative algebra (with identity) containing L as a vector subspace and generated by L .

ii) any linear map λ of L into an associative algebra A with $\lambda(x)\lambda(y) = \lambda(y)\lambda(x)$ all $x,y \in L$, extends to a (unique) homomorphism $SL \longrightarrow A$.

If $\lambda: L \longrightarrow L$ is any linear map then the automorphism λ_a and the derivation λ_d of TL leave the ideal I invariant and therefore induce an automorphism and a derivation of SL.

If $x_1,\ldots,x_n$ is a basis for L then

$$SL \approx \mathbb{R}[x_1,\ldots,x_n],$$

the algebra of polynomials over $x_1,\ldots,x_n$. With this identification, the extension to SL of the linear map $L \longrightarrow \mathbb{R}$, $x_i \longmapsto a_i$, is

$$f(x_1,\ldots,x_n) \longmapsto f(a_1,\ldots,a_n).$$

See [10] V. 18., 19. for details.

<u>Automorphisms and derivations.</u>

Let A be a, not necessarily associative, finite dimensional real algebra. Then the Lie group $\mathrm{Aut}(A)$ of automorphisms of A has Lie algebra $D(A)$ the algebra of derivations of A. To see this we note that if D is any derivation of A then from $D(xy) = (Dx)y + x(Dy)$ we have by induction

$$D^k(xy) = \sum_{i+j=k} \frac{k!}{i!\,j!} \cdot D^i x \, D^j y .$$

Therefore

$$e^{tD}(xy) = \sum_k \frac{t^k}{k!} \cdot D^k xy = \sum_i \sum_j \frac{t^i D^i x}{i!} \cdot \frac{t^j D^j y}{j!}$$

$$= (e^{tD}x)(e^{tD}y) \quad ,$$

so that $\exp tD \in \mathrm{Aut}(A)$ all $t \in \mathbb{R}$.

Conversely, if

$$e^{tD}(xy) = (e^{tD}x)(e^{tD}y)$$

all $t \in \mathbb{R}$ then

$$xy + tD(xy) + \frac{t^2}{2!} D^2(xy) + \ldots\ldots$$

$$= (x + tDx + \frac{t^2}{2!} D^2x + \ldots)(y + tDy + \frac{t^2}{2!} D^2y + \ldots)$$

all $t \in \mathbb{R}$, so that

$$D(xy) = (Dx)y + x(Dy)$$

and D is a derivation.

Action of change of observer.

Let $G = M \circledS \mathcal{L}_0$ be the connected group of restricted relativistic changes of observer. Let

$$\dot{T}: LG \longrightarrow S(D_T)$$

be the Lie algebra homomorphism defined by the requirement of relativistic invariance, see sections 3 and 4. By the universal property $\dot{T}$ extends to a unique homomorphism of associative algebras

$$\dot{T}: ULG \longrightarrow Op(D_T)$$

of ULG into the associative algebra $Op(D_T)$ generated by $S(D_T)$. For each $u \in ULG$, $\dot{T}(u)$ is an operator with domain D_T and is a sum of products of operators $\dot{T}(X)$, $X \in LG$. A change $g \in G$ of observer induces an automorphism $Ad_g: LG \longrightarrow LG$ of relativistic observables, which extends to a unique automorphism of ULG which we also denote by Ad_g The physical interpretation of this is that the operator $\dot{T}(Ad_g(u))$ plays the same role in the eyes of the new observer as $\dot{T}(u)$ did for the old. The map

$$Ad : G \longrightarrow Aut\ (ULG)$$

gives the action of G on ULG .

An element $u \in ULG$ is called an <u>invariant of G</u> if $Ad_g(u) = u$ for all $g \in G$. In this case the operator $\dot{T}(u)$ has a physical significance independent of a choice of relativistic observer.

Characterisation of Invariants.

Let LG be the Lie algebra of a connected Lie group G . Let TLG be the tensor algebra over LG , STLG the vector subspace of symmetric tensors, SLG the symmetric algebra, ULG the universal enveloping algebra. We have a diagram

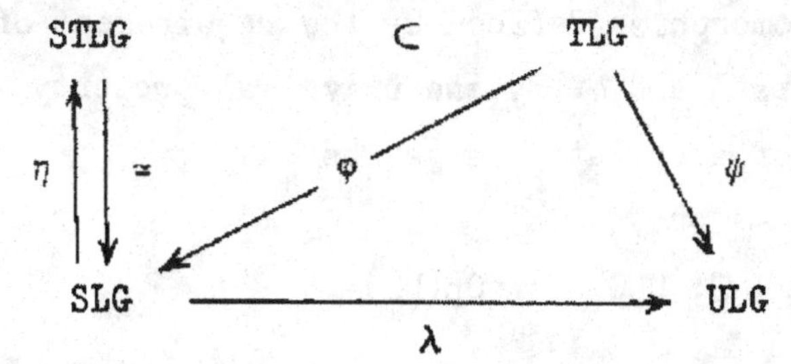

where φ, ψ are the canonical surjections. By [9] III § 5 Proposition 6 φ maps STLG bijectively to SLG , with inverse η say. We write $\lambda = \psi\eta$. By a dimension argument the linear map λ is bijective. See [9] V. § 6 proposition 2 .

The automorphism Ad_g of LG induces a unique automorphism of the tensor algebra TLG , which induces automorphisms of ULG and SLG also denoted by Ad_g .

The Lie group homomorphism $Ad : G \longrightarrow Aut\,(LG)$ induces a Lie algebra homomorphism

$$ad : LG \longrightarrow D(LG) ,$$

called the adjoint representation of LG . The derivation $ad_X : LG \longrightarrow LG$ is in fact the map $ad_X(Y) = [X,Y]$; see [7] IV § XI for example.

The derivation ad_X of LG induces a unique derivation of the tensor algebra TLG, which induces derivations of ULG and SLG also denoted by ad_X.

It follows that the diagrams

$$\begin{array}{ccc} SLG & \xrightarrow{\lambda} & ULG \\ {\scriptstyle Ad_g}\downarrow & & \downarrow{\scriptstyle Ad_g} \\ SLG & \xrightarrow[\lambda]{} & ULG \end{array} \qquad\qquad \begin{array}{ccc} SLG & \xrightarrow{\lambda} & ULG \\ {\scriptstyle ad_X}\downarrow & & \downarrow{\scriptstyle ad_X} \\ SLG & \xrightarrow[\lambda]{} & ULG \end{array}$$

commute, for each $g \in G$ and $X \in LG$. The invariants of G therefore correspond under λ to the elements of SLG invariant under Ad_G:

$$\{ s \in SLG \mid Ad_g(s) = s \text{ , all } g \in G \} .$$

<u>Theorem.</u> <u>The set of invariants of a connected Lie group</u> G <u>equals the centre of</u> ULG.

<u>Proof.</u> ULG has a natural filtration by finite dimensional subspaces

$$U_1LG \subset \ldots\ldots \subset U_rLG \subset \ldots\ldots$$

where U_rLG is the vector subspace spanned by products of the form

$$X_{i_1} X_{i_2} \ldots\ldots\ldots X_{i_r} ,$$

$X_i \in LG$. For each $X \in LG$ the derivation ad_X of ULG leaves the finite dimensional subspace U_rLG invariant, so that the restriction to U_rLG of the series

$$1 + ad_X + \frac{(ad_X)^2}{2!} + \frac{(ad_X)^3}{3!} + \ldots$$

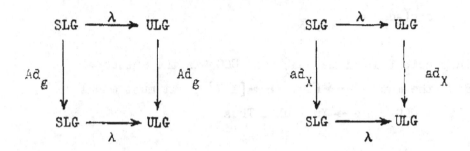

converges. It follows that the series converges on ULG, and defines an endomorphism $\exp(ad_X)$ of ULG since ad_X is a derivation. The restriction of $\exp(ad_X)$ to LG equals $Ad_{\exp X}$ by [7] IV § IX proposition 1, and therefore by uniqueness of the extension to ULG,

$$\exp(ad_X) = Ad_{\exp X}$$

on ULG. We further note that since ad_X on ULG is the unique derivation extending the map $LG \longrightarrow LG$, $Y \longmapsto [X,Y]$, it must equal the derivation $ULG \longrightarrow ULG$, $u \longmapsto Xu - uX$. Thus

$$ad_X(u) = Xu - uX$$

for all $u \in ULG$.

Since G is connected it is generated by the set of elements $\exp X$, $X \in LG$. Therefore $u \in ULG$ is invariant

iff $Ad_{\exp tX}(u) = u$, all $X \in LG$, $t \in \mathbb{R}$

iff $(\exp \circ ad_{tX})(u) = u$, " "

iff $u + t(ad_X)u + \frac{t^2}{2!}\cdot(ad_X)^2u + \dots = u$

iff $ad_X(u) = 0$, all $X \in LG$

iff $Xu - uX = 0$, "

iff $u \in$ centre ULG.

This completes the proof.

Invariants of the inhomogeneous Lorentz group.

With the notation of section 4, $(\pi_1, \pi_2, \pi_3, \pi_4)$ is an orthonormal basis of the Minkowski space $LM \approx M$, and the action of $\mathcal{L}$ on LM in the adjoint representation is the same as the natural action of $\mathcal{L}$ on M. The Lorentz scalar product gives isomorphisms

$$\otimes^2 M \approx \mathrm{Hom}(M,M) \approx (\otimes^2 M)^*$$

which commute with that action of $\mathcal{L}$, where $*$ denotes the vector space dual. These isomorphisms are characterised by

$$x \otimes y \longmapsto (\, z \longmapsto \langle y, z \rangle x)$$

and

$$A \longmapsto (\, x \otimes y \longmapsto \langle x, Ay \rangle)$$

respectively. We identify these spaces, (this being the customary identification of contravariant, mixed, and covariant tensors) and note that the skew-symmetric tensors in $\otimes^2 M$ correspond to $L\mathcal{L} \subset \mathrm{Hom}(M,M)$. Indeed $\pi_i \otimes \pi_j - \pi_j \otimes \pi_i$ corresponds to m_{ij}. The scalar product $\langle \cdot\cdot, \cdot\cdot \rangle$ is an element of $(\otimes^2 M)^*$ invariant under $\mathcal{L}$ and corresponds to:

$$\sum_{i,j} \langle \pi_i, \pi_j \rangle \, \pi_i \otimes \pi_j = -\pi_1 \otimes \pi_1 - \pi_2 \otimes \pi_2 - \pi_3 \otimes \pi_3 + \pi_4 \otimes \pi_4$$

in $\otimes^2 M$. It follows that the corresponding element of $UL\mathcal{L}$:

$$-\pi_1^2 - \pi_2^2 - \pi_3^2 + \pi_4^2$$

is an invariant of $\mathcal{L}$.

Let $w_i = \frac{1}{2.3!} \cdot \sum_{j,k,l} \varepsilon_{ijkl}\, \pi_j \otimes m_{kl}$ in $\otimes^3 M$, where ε_{ijkl} is completely skew-symmetric and $\varepsilon_{1234} = 1$. Then :

$$w_i = \frac{1}{2.3!} \cdot \sum_{j.k,l} \varepsilon_{ijkl}\, \pi_j \otimes (\pi_k \otimes \pi_l - \pi_l \otimes \pi_k)$$

$$= \frac{1}{3!} \cdot \sum_{j,k,l} \varepsilon_{ijkl} \cdot \pi_j \otimes \pi_k \otimes \pi_l$$

$$= \pi_j \wedge \pi_k \wedge \pi_l$$

where i. j. k. l is any even permutation of 1, 2, 3, 4 . The Lorentz scalar product and the basis $\pi_1, \pi_2, \pi_3, \pi_4$ defines an isomorphism $\theta : \Lambda^3 M \longrightarrow M$, characterised by

$$w \wedge x = \langle \theta(w), x \rangle\, \pi_1 \wedge \pi_2 \wedge \pi_3 \wedge \pi_4$$

for all $w \in \Lambda^3 M$ and $x \in M$. Moreover θ commutes with all Lorentz transformations of determinant 1 . Under this isomorphism, w_1, w_2, w_3, w_4 correspond to $-\pi_1, -\pi_2, -\pi_3, \pi_4$ respectively, Therefore

$$-w_1 \otimes w_1 - w_2 \otimes w_2 - w_3 \otimes w_3 + w_4 \otimes w_4$$

in $\Lambda^3 M \otimes \Lambda^3 M$ is invariant under $\mathcal{L}_0$. It follows that the corresponding element of $UL\mathcal{L}$:

$$-w_1^2 - w_2^2 - w_3^2 + w_4^2$$

is an invariant of $\mathcal{L}$, where here w_i denotes the element

$$\frac{1}{2.3!} \cdot \sum_{j.k.l} \varepsilon_{ijkl} \cdot \pi_j m_{kl} \quad \text{in } UL\mathcal{L}.$$

The unquantised relativistic relation between the linear momentum (p_1, p_2, p_3), the energy E, and the rest mass m of a particle is: $-p_1^2 - p_2^2 - p_3^2 + E^2 = m^2$. We therefore associate the invariant $-\pi_1^2 - \pi_2^2 - \pi_3^2 + \pi_4^2$ with the square of the mass. More specifically, if T is the representation of $M \circledS \mathcal{L}_0$ in the Hilbert space H associated with the states of a quantum mechanical system, and if $P_j = \frac{1}{i}\dot{T}(\pi_j)$ are the self-adjoint operators representing linear momentum and energy, then the self-adjoint positive square root (if any) of the operator $-P_1^2 - P_2^2 - P_3^2 + P_4^2$ represents the mass of the system.

Determination of invariants

We now indicate how invariants of a semi-simple Lie group G may be constructed. Take any finite dimensional, not necessarily unitary, representation T of G with representation space V. For each $g \in G$ the commutative diagram

$$\begin{array}{ccc} G & \xrightarrow{\;T\;} & GL(V) \\ {\scriptstyle I(g)}\downarrow & & \downarrow{\scriptstyle T(g)(\cdot)\,T(g^{-1})} \\ G & \xrightarrow{\;T\;} & GL(V) \end{array}$$

gives a commutative diagram

$$\begin{array}{ccccc} LG & \xrightarrow{\;\dot{T}\;} & Hom(V) & & \\ & & & \searrow{\scriptstyle \det_t} & \\ {\scriptstyle Ad_g}\downarrow & & \downarrow{\scriptstyle T(g)(\cdot)\,T(g^{-1})} & & \mathbb{R} \\ & & & \nearrow{\scriptstyle \det_t} & \\ LG & \xrightarrow{\;\dot{T}\;} & Hom(V) & & \end{array}$$

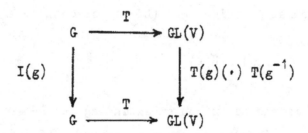

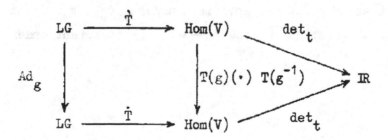

where $\det_t A = \det(A - t1)$ for each $t \in \mathbb{R}$. We will have

$$\det(\dot{T}(X) - t1) = \sum_i Q_i(X) t^i$$

say, where Q_i is a polynomial function on LG and

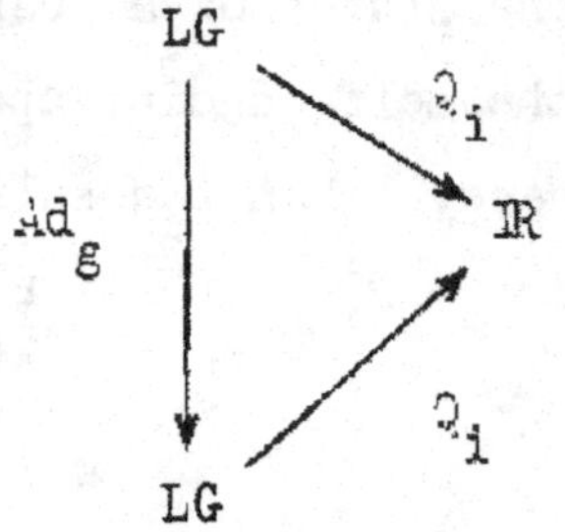

is commutative. This shows that Q_i is a polynomial function on LG invariant under AdG . We now establish a relation between such polynomial functions and the invariants of G .

Let $(LG)^*$ be the vector space dual of LG . There is a natural isomorphism $P \mapsto \tilde{P}$ from the algebra SLG of polynomials over LG to the algebra of polynomial functions on $(LG)^*$: for each $u \in (LG)^*$ define

$$\tilde{P}(u) = \tilde{u}(P)$$

where $\tilde{u}\colon SLG \longrightarrow \mathbb{R}$ is the homomorphism extending $u\colon LG \longrightarrow \mathbb{R}$. Each linear map $A\colon LG \longrightarrow LG$ induces an endomorphism of SLG . The transpose A^* induces an endomorphism of the algebra of polynomial functions on $(LG)^*$. The map $P \mapsto \tilde{P}$ commutes with these endomorphisms. These considerations show that $P \in SLG$ is invariant under Ad_g , $g \in G$, if and only if the polynomial function $\tilde{P}\colon (LG)^* \longrightarrow \mathbb{R}$ is invariant under $(Ad_g)^*$.

The Killing form on LG : $< X,Y > = \text{trace}\,(ad_X \circ ad_Y)$ defines a linear map α: $LG \longrightarrow (LG)^*$, $\alpha(X)Y = < X,Y >$. If A is any automorphism of LG then the diagram

$$\begin{array}{ccc} LG & \xrightarrow{\alpha} & (LG)^* \\ {\scriptstyle A}\downarrow & & \downarrow{\scriptstyle (A^{-1})^*} \\ LG & \xrightarrow{\alpha} & (LG)^* \end{array}$$

is commutative since

$$\alpha(AX)Y = <AX,Y> = <X,A^{-1}Y> = \alpha(X)A^{-1}Y = [(A^{-1})^*\alpha](X)Y .$$

We have here used the fact that the Killing form is invariant under all automorphisms. We have now shown that a polynomial function $\tilde{P}$ on $(LG)^*$ invariant under $(Ad_g{}^{-1})^*$ defines a polynomial function $\tilde{P}\alpha$ on LG invariant under Ad_g , $g \in G$.

Since G is semi-simple the Killing form is non-singular and α is bijective. It follows that every polynomial function on LG invariant under Ad_g gives rise to a polynomial function on $(LG)^*$ invariant under $(AdG)^*$ and hence to an element of SLG invariant under AdG , and finally to an invariant of G .

<u>Note.</u> Let $P = P(X_1,\ldots,X_n) \in SLG = \mathbb{R}[X_1,\ldots,X_n]$ be a polynomial over LG , expressed as a polynomial in the basis elements $X_1,\ldots,X_n$.

The corresponding polynomial function $Q = \tilde{P}\alpha$ on LG is given by

$$Q(X) = \tilde{P}(\alpha(X)) = \widetilde{\alpha(X)}(P)$$

$$= P(\alpha(X)X_1,\dots,\alpha(X)X_n)$$

$$= P(\langle X,X_1\rangle,\dots,\langle X,X_n\rangle)\ .$$

__Example.__

$G = SO(3)$, the rotation group of $\mathbb{R}^3$.

$LG = so(3)$, skew-symmetric matrices.

The basis

$$X_1 = \begin{pmatrix} 0 & 1 & 0 \\ -1 & 0 & 0 \\ 0 & 0 & 0 \end{pmatrix}, \quad X_2 = \begin{pmatrix} 0 & 0 & 1 \\ 0 & 0 & 0 \\ -1 & 0 & 0 \end{pmatrix}, \quad X_3 = \begin{pmatrix} 0 & 0 & 0 \\ 0 & 0 & 1 \\ 0 & -1 & 0 \end{pmatrix}$$

is orthonormal with respect to the Killing form. The identity representation gives

$$LG \longrightarrow so(3) \xrightarrow{\det_t} \mathbb{R}$$

$$a_1X_1 + a_2X_2 + a_3X_3 \longmapsto \det \begin{pmatrix} -t & a_1 & a_2 \\ -a_1 & -t & a_3 \\ -a_2 & -a_3 & -t \end{pmatrix} = -t^3 - (a_1^2 + a_2^2 + a_3^2)\cdot t\ .$$

Thus

$$Q(a_1X_1 + a_2X_2 + a_3X_3) = a_1^2 + a_2^2 + a_3^2$$

is a polynomial function on LG invariant under AdG . The corresponding polynomial $P \in SLG$ is given by

$$a_1^2 + a_2^2 + a_3^2 = Q(X) = P(\langle X,X_1\rangle, \langle X,X_2\rangle, \langle X,X_3\rangle)$$

$$= P\langle -2a_1 , -2a_2 , -2a_3 \rangle$$

so that $P = \frac{1}{4}(X_1^2 + X_2^2 + X_3^2) \in SLG$. The corresponding invariant of G is

$$\lambda(P) = \frac{1}{4}(X_1^2 + X_2^2 + X_3^2) \in ULG .$$

This invariant, however, can be obtained much more directly by the method used earlier in constructing invariants of the inhomogeneous Lorentz group .

Section 6. Induced Representations.

Let G be a locally compact group, and K a closed subgroup. Let $\sigma : K \longrightarrow U(V)$ be a unitary representation of K on a Hilbert space V. We will now describe a process whereby a unitary representation of the whole group G is defined.

On the topological product $G \times V$ define an equivalence relation

$$(gk,v) \sim (g, \sigma(k)v)$$

for all $k \in K$. Let

$$G \times_K V = \{ [g,v] \mid g \in G , v \in V \}$$

be the quotient topological space, where $[g,v]$ is the equivalence class of (g,v). Let

$$\pi : G \times_K V \longrightarrow G/K$$

be the map $\pi[g,v] = gK$.

For any $p \in G/K$ the inverse image $\pi^{-1}(p)$ has a natural Hilbert space structure. Indeed if $p = gK$ then each element of $\pi^{-1}(p)$ is of the form $[g,v]$ with $v \in V$ uniquely determined. The bijection $[g,v] \longmapsto v$ of $\pi^{-1}(p)$ onto V gives $\pi^{-1}(p)$ the structure of a Hilbert space. This does not depend on the choice of g with $p = gK$ since if $p = gK = g_1K$ then the diagram

$$\begin{array}{ccc} & [g,v] \longmapsto v & \\ & \nearrow & V \\ \pi^{-1}(p) & & \Big\downarrow \sigma(g_1^{-1}\cdot g) \\ & \searrow & V \\ & [g_1,v] \longmapsto v & \end{array}$$

commutes. Since $\sigma(g_1^{-1}\cdot g)$ is unitary our assertion follows.

Hilbert bundles.

The considerations above suggest the following definition. A triple $\xi = (X,\pi,Y)$ is a Hilbert bundle if X and Y are topological spaces, π a continuous surjection of X on Y, and $\pi^{-1}(p)$ is given a Hilbert space structure for each $p \in Y$. X is then called the total space of the bundle ξ, π the projection, and Y the base space. The Hilbert space $\pi^{-1}(p)$ is the fibre over p. A section ψ of the bundle is a map $\psi: Y \longrightarrow X$ such that $\pi\psi = 1_Y$. Thus a section is a function ψ on the base space Y with its value $\psi(p)$ in the fibre $\pi^{-1}(p)$ for each p.

Let $\xi = (X,\pi,Y)$ and $\xi_1 = (X_1,\pi_1,Y_1)$ be Hilbert bundles. A Hilbert bundle isomorphism $\xi \longrightarrow \xi_1$ is a pair (α,β) of homeomorphisms $\alpha: X \longrightarrow X_1$ and $\beta : Y \longrightarrow Y_1$ such that

i) $\pi_1\alpha = \beta\pi$

ii) α maps the Hilbert space $\pi^{-1}(p)$ isometrically onto $\pi_1^{-1}(\beta p)$ for each $p \in Y$.

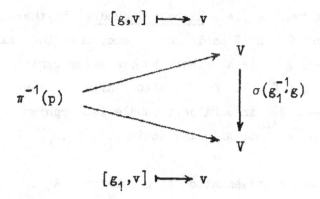

Let G be a topological group. ξ is a <u>Hilbert G-bundle</u> if there are given continuous actions of G on X and on Y such that the pair $\alpha_g : x \mapsto g\,x$ and $\beta_g : y \mapsto g\,y$ is a Hilbert bundle automorphism (α_g, β_g) of ξ for each $g \in G$. If ξ_1 is also a Hilbert G-bundle then a <u>G-isomorphism</u> $\xi \longrightarrow \xi_1$ is a Hilbert bundle isomorphism (α,β) such that α and β commute with the action of G .

If σ is a unitary representation of a subgroup K of G in a Hilbert space V , then the triple

$$\xi_\sigma = (G \times_K V\ ,\ \pi\ ,\ G/K)$$

introduced above is a Hilbert G-bundle relative to the actions

$$g[g_1,v_1] = [gg_1,v_1]$$

and

$$g(g_1K) = gg_1K$$

of G on $G \times_K V$ and G/K respectively. If τ is a representation of K in W and σ and τ are equivalent representations, with the equivalence given by an isometry

$$A: V \longrightarrow W\ ,$$

then ξ_σ and ξ_τ are G-isomorphic under the map

$$[g,v] \longmapsto [g,Av]$$

and the identity map on the base space G/K .

Measure theoretical notions.

Let X be a topological space. The class of Borel sets of X is the smallest family of subsets of X which includes the open sets and which is closed under i) complements ii) countable unions. If Y is a topological space then a Borel function $\psi: X \longrightarrow Y$ is a map for which $\psi^{-1}(E)$ is Borel in X whenever E is Borel in Y. A measure in X is a function μ on the class of Borel sets of X with values in $[0,\infty]$ such that

$$\text{i)} \quad \mu\Big(\bigcup_{i=1}^{\infty} E_i\Big) = \sum_{i=1}^{\infty} \mu(E_i)$$

for each countable sequence $\{E_i\}$ of mutually disjoint Borel sets.

ii) $\mu(E) < \infty$ for each compact Borel set E.

Two measures μ,μ on X are equivalent if they assume the value zero for the same Borel sets in X. The measure class of μ is its equivalence class under this equivalence relation. For any two measures μ,ν in the same measure class there exists a unique non-negative Borel function on X, denoted by $\frac{d\nu}{d\mu}$ and called the Radon-Nikodym derivative of ν with respect to μ, such that

$$\int_X f(p)d\nu(p) = \int_X f(p)\frac{d\nu}{d\mu}(p)d\mu(p)$$

for each integrable complex Borel function f on X.

Now let K be a closed subgroup of a locally compact group $\mathfrak{G}$. There is a unique non-zero measure class M in G/K such that for each μ in M and each $g \in G$ the measure μ_g:

$$\mu_g(E) = \mu(g^{-1}E)$$

is also in M . M is called the invariant measure class on G/K . See [5] § 2, no 5 .

Induced representations.

Let $\xi = (X, \pi, Y)$ be a Hilbert G-bundle, where G is a topological group, and let M be an invariant measure class on Y . For each $\mu \in M$ let

$$\mathcal{H}_\mu = \{ \psi | \psi \text{ a Borel section of } \xi \text{ , and } \int_Y \langle \psi(p), \psi(p) \rangle \, d\mu(p) < \infty \}$$

where $\langle \psi(p), \psi(p) \rangle$ denotes the inner product in the Hilbert space $\pi^{-1}(p)$. With the inner product

$$\langle \psi, \varphi \rangle = \int_Y \langle \psi(p) , \varphi(p) \rangle \, d\mu(p)$$

and with sections identified if they differ only on a set of μ-measure zero, $\mathcal{H}_\mu$ is a Hilbert space.

Define an action of G on $\mathcal{H}_\mu$ by

$$(g\,\psi)(p) = \sqrt{\frac{d\mu_g}{d\mu}}(p)\, g(\psi(g^{-1} \cdot p))$$

for each $p \in Y$. This is the usual rule for shifting functions on Y apart from the weighting factor $\sqrt{\frac{d\mu_g}{d\mu}}$. It is a unitary action since

$$\langle g\,\psi,\ g\,\varphi\rangle = \int_Y \frac{d\mu_g}{d\mu}\ \langle g\,(\psi(g^{-1}\cdot p)),\ g(\varphi(g^{-1}\cdot p))\rangle\ d\mu(p)$$

$$= \int_Y \ \langle \psi(g^{-1}\cdot p),\ \varphi(g^{-1}\cdot p)\rangle\ d\mu(g^{-1}\cdot p)$$

$$= \ \langle \psi,\ \varphi \rangle\ .$$

It can be shown that this action of G on $\mathcal{H}_\mu$ gives a (strongly continuous) representation of G . We denote this representation by $T_\mu(\xi)$. If ν is any other measure in the measure class of M then $\psi \longmapsto \sqrt{\frac{d\mu}{d\nu}}\cdot\psi$ defines an isometry $\mathcal{H}_\mu \longrightarrow \mathcal{H}_\nu$ which gives an equivalence between $T_\mu(\xi)$ and $T_\nu(\xi)$. The equivalence class of $T_\mu(\xi)$ is therefore independent of the choice of μ in the invariant measure class M . If Y has a unique invariant measure class then we denote the representation of G simply by $T(\xi)$. We note that G-isomorphic bundles give equivalent representations.

When $\xi_\sigma = (G \times_K V,\ \pi,\ G/K)$ is the Hilbert G-bundle defined by a locally compact group G and a representation σ of a closed subgroup in a Hilbert space V , and the unique invariant measure class is taken on the base space G/K , then the representation $T(\xi_\sigma)$ is called the representation of G induced by σ . If σ and τ are equivalent representations of K then the bundles ξ_σ and ξ_τ are G-isomorphic and yield equivalent induced representations $T(\xi_\sigma)$ and $T(\xi_\tau)$.

For more material on induced representations see [22], [23], [24] and [25] .

Section 7. Representations of Semi-direct Products.

Let G be a separable locally compact group with an abelian normal subgroup N . Let H be another subgroup such that the multiplication map $N \times H \longrightarrow G$ is bijective. Then the equation

$$(n_1h_1)(n_2h_2) = n_1(h_1n_2h_1^{-1})\ h_1h_2$$

shows that $nh \mapsto (n,h)$ is an isomorphism of G onto the semi-direct product $N \circledS H$ relative to the action $n \mapsto hnh^{-1}$ of H on N .

A <u>character</u> χ of N is a continuous homomorphism

$$\chi \colon N \longrightarrow U(1)$$

of N into the complex numbers of absolute value 1 . The set of characters form an abelian group $\hat{N}$ with group operation given by $(\chi_1\chi_2)(n) = \chi_1(n)\ \chi_2(n)$. With respect to a suitable topology (the compact-open) $\hat{N}$ is a separable locally compact group called the <u>dual</u> of N . See [16] § 23 for details. Any bijective map $\alpha \colon N \longrightarrow N$ induces a map $\hat{N} \longrightarrow \hat{N}$, $\chi \longmapsto \alpha\chi$, defined by

$$(\alpha\chi)(n) = \chi(\alpha^{-1}n) .$$

In this way ~~the~~ an action $n \mapsto gng^{-1}$ of G on N induces an action $\chi \longmapsto g\chi$ of G on $\hat{N}$, $(g\chi)(n) = \chi(g^{-1}ng)$. For each $\chi \in \hat{N}$ we write

$$G\chi = \{\ g\chi \mid g \in G\ \}$$

for the orbit of χ under this action of G on $\hat{N}$; so that $\hat{N}$ is a disjoint union of orbits. We denote by

$$G_\chi = \{ g \mid g \in G \text{ and } g\chi = \chi \}$$

the isotropy group of χ under the action of G . Since N is abelian N acts trivially on N and hence on $\hat{N}$, so that N is a subgroup of G_χ . Let $L_\chi = H \cap G_\chi$, then

$$G_\chi = N \circledS L_\chi$$

is a semi-direct product. L_χ is called the little group of χ ; it is the isotropy group of χ under H . There is a natural bijection

$$G/G_\chi \longrightarrow G\chi$$

$gG_\chi \longmapsto g\chi$, which can be shown to preserve Borel sets, and which commutes with the action of G . This bijection will even be a homeomorphism under quite general conditions; see [5] Appendice I . We will in future often identify G/G_χ and $G\chi$.

Bundles over an orbit.

Let σ be a unitary representation of the little group L_χ with representation space. V . Then

$$\chi\sigma : (n,\ell) \longmapsto \chi(n)\sigma(\ell)$$

is a unitary representation of $G_\chi = N \circledS L_\chi$ in V .

Let $\xi_\sigma^\chi = (G \times_{G_\chi} V, \pi, G/_{G_\chi})$ be the Hilbert G-bundle, with base space $G/G_\chi \cong G\chi$, defined by the representation $\chi\sigma$. Thus for each representation σ of the little group L_χ we have a G-bundle ξ_σ^χ over the orbit $G\chi$.

The bundle ξ_σ^χ depends, up to G-isomorphism, only on the orbit $G\chi$ and not on the choice of χ. Indeed, if $G\chi = G\chi_1$, then $\chi_1 = g_1\chi$ with $g_1 \in G$. The little group of χ_1 is

$$L_{\chi_1} = g_1 L_\chi g_1^{-1} .$$

Let σ_1 be the representation of L_{χ_1} on V given by

$$\sigma_1(l) = \sigma(g_1^{-1} l g_1) .$$

The bundles ξ_σ^χ and $\xi_{\sigma_1}^{\chi_1}$ are then G-isomorphic under the maps

$$G \times_{G_\chi} V \longrightarrow G \times_{G_{\chi_1}} V , \quad [g,v] \longmapsto [g_1 g g_1^{-1}, v]$$

and

$$G\chi \longrightarrow G\chi_1 , \quad p \longmapsto g_1 p .$$

Theorem of Mackey. Let $G = N \circledS H$ be a semi-direct product satisfying the conditions above, and suppose that $\hat{N}$ contains a Borel subset with meets each orbit in $\hat{N}$ under G in just one point. Then

i) $T(\xi_\sigma^\chi)$ is an irreducible representation of G for each $\chi \in \hat{N}$ and each irreducible representation σ of L_χ.

ii) each irreducible representation of G is equivalent to one of the form $T(\xi_{\sigma}^{\chi})$ with the orbit $G\chi$ uniquely determined and σ determined up to equivalence.

We shall call a semi-direct product regular if it satisfies the conditions of Mackey's theorem. See [22] Theorem 14.2 for a proof of this theorem, or [24] Theorem 3.12 for a more general result.

Action of N .

The induced representation $T(\xi_{\sigma}^{\chi})$ takes a particularly simple form when restricted to the subgroup N. Let $\psi: G\chi \longrightarrow G \times_{G_{\chi}} V$ be a section of the bundle ξ_{σ}^{χ}, and let $n \in N$. For $p \in G\chi$ we have $p = g\chi$ say, so that $\psi(p) = [g,v]$ for some $v \in V$. Then

$$\begin{aligned}
(n\psi)(p) &= n(\psi(p)) \quad \text{since } N \text{ acts trivially on } G\chi \subset \hat{N} \\
&= n[g,v] \\
&= [gg^{-1}ng,v] \\
&= [g,\ \chi(g^{-1}ng)v] \\
&= [g,\ (g\chi)(n)v] \\
&= p(n)[g,v] \\
&= n(p)\ \psi(p)
\end{aligned}$$

where $n \in N$ is considered as a function on $\hat{N}$ defined by the rule $n(p) = p(n)$.

Thus the operation of n on the section ψ is simply multiplication of the section by the function $p \longmapsto n(p)$ on the base space $G\chi$.

Action of H.

Since N acts trivially on $\hat{N}$, the map

$$H/L_\chi \longrightarrow G\chi \ , \qquad hL_\chi \longmapsto h\chi$$

is bijective and we shall identify these sets. Let ω be a fixed section of the bundle $H \longrightarrow H/L_\chi$. This means that ω is a map $G\chi \longrightarrow H$ such that

$$\omega(p)\chi = p$$

for each $p \in G\chi$. Each element in the fibre $\pi^{-1}(p)$ of the bundle ξ^χ_σ has a unique expression in the form

$$[\omega(p), v]$$

with $v \in V$. It follows that any section ψ of this bundle has a unique expression in the form

$$\psi(p) = [\omega(p), \psi_\omega(p)]$$

where ψ_ω is a function on $G\chi$ with values in V.

For simplicity we assume now that the action of G on the sections of ξ_σ^χ is relative to a measure μ on $G\chi$ which is invariant under G :

$$\frac{d\mu_g}{d\mu} = 1 \ .$$

For each $h \in H$ we then have

$$(h\psi)(p) = h(\psi(h^{-1}p)) = h[\omega(h^{-1}p),\psi_\omega(h^{-1}p)]$$

$$= [\omega(p)\ \omega(p)^{-1}h\omega(h^{-1}p),\psi_\omega(h^{-1}p)]$$

$$= [\omega(p),\sigma(\omega(p)^{-1}h\omega(h^{-1}p))\psi_\omega(h^{-1}p)] \ .$$

Thus the induced action of H on ψ_ω is:

$$(h\psi_\omega)(p) = \sigma(\omega(p)^{-1}h\omega(h^{-1}p))\psi_\omega(h^{-1}p)$$

Let $\mathcal{H}$ be the Hilbert space of square integrable Borel sections of the bundle ξ_σ^χ . The linear map

$$\psi \longmapsto \psi_\omega$$

gives a Hilbert space structure to the vector space

$$\mathcal{H}_\omega = \{ \psi_\omega \mid \psi \in \mathcal{H} \} \ ,$$

which consists of certain functions on the orbit $G\chi$ with values in V .

The inner product in $\mathcal{H}_\omega$ is given by

$$\langle\psi_\omega,\varphi_\omega\rangle = \langle\psi,\varphi\rangle = \int_{G\chi} \langle\psi(p),\varphi(p)\rangle \, d\mu(p)$$

$$= \int_{G\chi} \langle\psi_\omega(p),\varphi_\omega(p)\rangle \, d\mu(p)$$

which is the usual inner product of functions on $G\chi$ with values in a Hilbert space V . We note however that ψ_ω will not in general be a Borel function unless ω is a Borel section.

Let T be the representation of H on the Hilbert space $\mathcal{H}_\omega$ defined above:

$$(T(h)\psi_\omega)(p) = \sigma(\omega(p)^{-1}h\omega(h^{-1}p))\psi_\omega(h^{-1}p) \ .$$

The group L_χ acts on the bundle $H \longrightarrow G\chi$ by

$$\begin{array}{ccccc} H & \longrightarrow & H & , & h \longmapsto lhl^{-1} \\ \omega \uparrow\downarrow & & \downarrow\uparrow \omega & & \\ G\chi & \longrightarrow & G\chi & , & p \longmapsto lp \end{array}$$

each $l \in L_\chi$. Let X belong to the Lie algebra of L_χ . The action of the operator $\dot{T}(X)$ on $\mathcal{H}_\omega$ is particularly simple when ω is an equivariant section of $H \longrightarrow G\chi$ relative to the action of the 1-parameter group $t \longmapsto \exp tX$. Indeed, in this case we have

$$\omega((\exp tX)p) = (\exp tX)\omega(p)(\exp -tX)$$

all $t \in \mathbb{R}$, so that

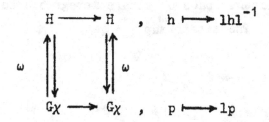

$$(T(\exp tX)\psi_\omega)(p) = \sigma(\exp tX)\psi_\omega((\exp -tX)p) .$$

Thus

$$(\dot{T}(X)\psi_\omega(p) = \frac{d}{dt}\{T(\exp tX)\psi_\omega\}(p)|_{t=0}$$

$$= \dot{\sigma}(X)(\psi_\omega(p)) + \frac{d}{dt}\psi_\omega((\exp -tX)p)|_{t=0} .$$

This means that $\dot{T}(X)$ is the sum of operators $\dot{T}_1(X)$ and $\dot{T}_2(X)$ on $\mathcal{H}_\omega$, where

$$\dot{T}_1(X)\psi_\omega = \dot{\sigma}(X)\psi_\omega$$

and

$$\dot{T}_2(X)\psi_\omega = \frac{d}{dt}\psi_\omega((\exp -tX)p)|_{t=0} .$$

We may note that

$$\psi_\omega \longmapsto \sigma(l)\psi_\omega$$

is a unitary operator on $\mathcal{H}_\omega$ for each $l \in L_\chi$. This implies in particular that $\dot{T}_1(X)$ is the generator of a 1-parameter group of unitary transformations and therefore skew adjoint by Stones theorem. $\dot{T}(X)$ is also skew adjoint by Stones theorem, so that $\dot{T}_2(X)$ is as well.

We note further that the operator $\frac{1}{i}\dot{T}_1(X)$ on $\mathcal{H}_\omega$ has the same spectrum as the operator $\frac{1}{i}\dot{\sigma}(X)$ on V .

Section 8. Classification of the Relativistic Free Particles.

In section 3 we have defined a quantum mechanical system to be an elementary relativistic free particle if it is associated with an irreducible representation of the covering group $M \circledS \tilde{\mathcal{L}}_o$ of the restricted inhomogeneous Lorentz group. A choice of relativistic observer $M \xrightarrow{\alpha} \mathbb{R}^4$ induces an isomorphism

$$\alpha_* : M \circledS \tilde{\mathcal{L}}_o \longrightarrow \mathbb{R}^4 \circledS SL(2,\mathbb{C})$$

so that, in order to classify the possible elementary relativistic free particles, we must determine the irreducible representations of $\mathbb{R}^4 \circledS SL(2,\mathbb{C})$. To do this we will apply the theorem of Mackey from the previous section.

Orbits and little groups.

Each $p \in \mathbb{R}^4$ defines a character

$$\chi_p : \mathbb{R}^4 \longrightarrow U(1)$$

where $\chi_p(x) = e^{i\langle p,x\rangle}$ and $\langle p,x\rangle$ denotes the Lorentz scalar product. The map $p \mapsto \chi_p$ is an isomorphism of $\mathbb{R}^4$ onto its character group $\hat{\mathbb{R}}^4$ as a topological group, see [16] (23.27). If $A: \mathbb{R}^4 \longrightarrow \mathbb{R}^4$ is a homogeneous Lorentz transformation then

$$\chi_{Ap}(x) = e^{i\langle Ap,x\rangle} = e^{i\langle p,A^{-1}x\rangle} = \chi_p(A^{-1}x) = A\chi_p(x)$$

so that $\chi_{Ap} = A\chi_p$. The isomorphism $\mathbb{R}^4 \approx \hat{\mathbb{R}}^4$ thus preserves the action of $SL(2,\mathbb{C})$. We will therefore identify $\mathbb{R}^4$ with its dual $\hat{\mathbb{R}}^4$ by the map $p \longmapsto \chi_p$. Thus

$$p(x) = x(p) = e^{i\langle p,x\rangle}$$

In applying the theorem of Mackey to determine the irreducible representations of $\mathbb{R}^4 \circledS SL(2,\mathbb{C})$ we have $G = \mathbb{R}^4 \circledS SL(2,\mathbb{C})$, $H = SL(2,\mathbb{C})$ and $N = \mathbb{R}^4 = \hat{N}$. The orbits in $\hat{N}$ under G are the orbits in $\hat{N}$ under H since N acts trivially. We must therefore determine the orbits in $\mathbb{R}^4$ under $SL(2,\mathbb{C})$.

The action of $SL(2,\mathbb{C})$ on $\mathbb{R}^4$ is via the covering map $SL(2,\mathbb{C}) \longrightarrow SO(3,1)$. The orbits of $\mathbb{R}^4$ under $SL(2,\mathbb{C})$ are therefore the orbits under $SO(3,1)$. These orbits are

$$M_+^c = \{ p \mid \langle p,p\rangle = c > 0 \ , \ p_4 > 0 \}$$

$$M_-^c = \{ p \mid \langle p,p\rangle = c > 0 \ , \ p_4 < 0 \}$$

$$M^{-c} = \{ p \mid \langle p,p\rangle = -c < 0 \}$$

$$M_+^0 = \{ p \mid \langle p,p\rangle = 0, \ p_4 > 0 \}$$

$$M_-^0 = \{ p \mid \langle p,p\rangle = 0, \ p_4 < 0 \}$$

$$\{0\} \ .$$

To see this we note that

i) any point in $\mathbb{R}^4$ can be mapped into the half-plane $\{ p \mid p_1 = p_2 = 0 , p_3 \geq 0 \}$ by a space rotation.

ii) The orbits under the group of pure Lorentz transformations

$$\begin{pmatrix} 1 & & & \\ & 1 & & \\ & & \cosh u & \sinh u \\ & & \sinh u & \cosh u \end{pmatrix}$$

$u \in \mathbb{R}$, in this half-plane, are the hyperbolas and straight lines and point:

$$-p_3^2 + p_4^2 = c > 0 \, , \; p_4 > 0$$

$$-p_3^2 + p_4^2 = c > 0 \, , \; p_4 < 0$$

$$-p_3^2 + p_4^2 = -c < 0$$

$$p_3 = p_4 > 0$$

$$-p_3 = p_4 < 0$$

$$\{0\} .$$

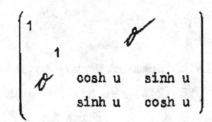

iii) the function $p \longmapsto \langle p,p \rangle$ is constant on each orbit

iv) each orbit is connected since $SL(2,\mathbb{C})$ is connected.

From properties i) and ii) we conclude that each of the sets in the family

$$\{M_+^c\} , \quad \{M_-^c\} , \quad \{M^{-c}\} \qquad\qquad (c > 0)$$

and M_+^o , M_-^o , $\{0\}$ are contained in one orbit. From iii) and iv) we conclude that no two of these sets is contained in the same orbit.

We note that $\mathbb{R}^4 Ⓢ SL(2,\mathbb{C})$ is a regular semi-direct product since the Borel set

$$\{(0,0,0,p_4) | p_4 \in \mathbb{R}\} \cup \{(0,0,p_3,0) | p_3 > 0 \} \cup \{(0,0,1,1)\} \cup \{(0,0,1,-1)\}$$

meets each orbit in just one point.

For each orbit $G\chi$ in $\mathbb{R}^4$, with representative χ, and each irreducible representation σ of the little group L_χ we have an irreducible representation $T(\xi_\sigma^\chi)$ of $\mathbb{R}^4 Ⓢ SL(2,\mathbb{C})$. The choice of representative point χ in each orbit, and the resulting little group, can be taken as follows.

Orbit	Representative	Little group
$G\chi$	χ	L_χ
M_+^c	$(0,0,0,\sqrt{c})$	$SU(2)$
M_-^c	$(0,0,0,-\sqrt{c})$	$SU(2)$
M^{-c}	$(0,\sqrt{c},0,0)$	$SL(2,\mathbb{R})$
M_+^o	$(0,0,1,1)$	Δ
M_-^o	$(0,0,1,-1)$	Δ
$\{0\}$	$(0,0,0,0)$	$SL(2,\mathbb{C})$

where

$$\Delta = \left\{ \begin{pmatrix} e^{i\theta} & z \\ o & e^{-i\theta} \end{pmatrix} \mid \theta \in \mathbb{R} , z \in \mathbb{C} \right\} .$$

These little groups are determined directly from the definition of the action of $SL(2,\mathbb{C})$ on $\mathbb{R}^4$. We have for instance

i) $x = (0,0,0, \sqrt{c})$ corresponds to the matrix $\underset{\sim}{x} = \begin{pmatrix} \sqrt{c} & o \\ o & \sqrt{c} \end{pmatrix}$ so that the little group is the subgroup of $SL(2,\mathbb{C})$:

$$\left\{ A \mid A\cdot \begin{pmatrix} \sqrt{c} & o \\ o & \sqrt{c} \end{pmatrix} \cdot A^* = \begin{pmatrix} \sqrt{c} & o \\ o & \sqrt{c} \end{pmatrix} \right\}$$

$$= \{ A \mid AA^* = 1 \}$$

$$= SU(2) .$$

ii) $x = (0,\sqrt{c},0,0)$ corresponds to $\underset{\sim}{x} = \begin{pmatrix} o & i\sqrt{c} \\ -i\sqrt{c} & o \end{pmatrix}$

so that the little group is

$$\left\{ A \mid A\begin{pmatrix} o & i\sqrt{c} \\ -i\sqrt{c} & o \end{pmatrix} A^* = \begin{pmatrix} o & i\sqrt{c} \\ -i\sqrt{c} & o \end{pmatrix} \right\}$$

$$= \left\{ A \mid A\begin{pmatrix} o & 1 \\ -1 & o \end{pmatrix} A^* = \begin{pmatrix} o & 1 \\ -1 & o \end{pmatrix} \right\}$$

$$= \left\{ A \mid A\begin{pmatrix} o & 1 \\ -1 & o \end{pmatrix} A^* = A\begin{pmatrix} o & 1 \\ -1 & o \end{pmatrix} A' \right\}$$

$$= \{ A \mid A^* = A' \}$$

$$= SL(2,\mathbb{C}) .$$

We have used the fact that $A\begin{pmatrix} o & 1 \\ -1 & o \end{pmatrix} A' = \begin{pmatrix} o & 1 \\ -1 & o \end{pmatrix}$ all $A \in SL(2,\mathbb{C})$.

iii) $x = (0,0,1,1)$ corresponds to $\underset{\sim}{x} = \begin{pmatrix} 2 & o \\ o & o \end{pmatrix}$

so that the little group is

$$\left\{ \begin{pmatrix} a & b \\ c & d \end{pmatrix} \mid \begin{pmatrix} a & b \\ c & d \end{pmatrix}\begin{pmatrix} 2 & o \\ o & o \end{pmatrix}\begin{pmatrix} \bar{a} & \bar{c} \\ \bar{b} & \bar{d} \end{pmatrix} = \begin{pmatrix} 2 & o \\ o & o \end{pmatrix} \right\}$$

$$= \left\{ \begin{pmatrix} a & b \\ c & d \end{pmatrix} \mid |a|^2 = 1 , c = o \right\}$$

$$= \Delta$$

Momentum, energy and mass.

Let $T = T(\xi_\sigma^\chi)$ be an irreducible representation of $G = \mathbb{R}^4 \circledS SL(2,\mathbb{C})$. In section 7 we determined the action $T(x)$ of $x \in \mathbb{R}^4$ on a section ψ of the Hilbert bundle ξ_σ^χ with base space the orbit $G\chi$. Indeed

$$(T(x)\psi)(p) = x(p)\psi(p) = e^{i\langle p,x\rangle}\psi(p)$$

so that

$$T(x) = e^{i\langle p,x\rangle}$$

where by this is understood the operator given by multiplication of sections by the function $p \mapsto e^{i\langle p,x\rangle}$ defined on the base space $G\chi$ of the bundle.

Let $\pi_1 \in L\mathbb{R}^4$ be the element corresponding to linear momentum in the x_1 direction, then

$$T(\exp t\pi_1) = T((t,0,0,0)) = e^{-itp_1}$$

so that

$$\dot{T}(\pi_1) = -ip_1$$

and

$$\frac{1}{i}\dot{T}(\pi_1) = -p_1 \ .$$

Thus the self-adjoint operator P_1 , which represents linear momentum in the x_1 direction, is simply multiplication by the function $-p_1$. Similarly for the other components of linear momentum. The operator $P_4 = \frac{1}{i}T(\pi_4)$ representing the energy is multiplication by the function p_4 The operator

$$-P_1^2 - P_2^2 - P_3^2 + P_4^2$$

is therefore multiplication by $\langle p,p \rangle$ which is a constant on the orbit $G\chi$. The system is therefore an eigenspace of the mass operator, and the mass is the constant value of $\sqrt{\langle p,p \rangle}$ on the orbit.

We can now write down the following table giving the spectrum of the energy operator, and the mas , of the particle associated with the representation $T(\xi_\sigma^\chi)$. We note that it depends only on the orbit $G\chi$ and not on σ .

Orbit	Energy spectrum	Mass
M_+^c	$[\sqrt{c},\infty)$	$\sqrt{c}$
M_-^c	$(-\infty,-\sqrt{c}]$	$\sqrt{c}$
M^{-c}	$(-\infty,\infty)$	$\sqrt{-c}$
M_+^0	$(0,\infty)$	0
M_-^0	$(-\infty,0)$	0
$\{0\}$	$\{0\}$	0

The case of imaginary mass is excluded since the mass operator would not be self-adjoint. The only particle with zero energy spectrum is the vacuum state. If we exclude negative energy states then we are left with the orbits M_+^c , $c > 0$, and M_+^o which we will refer to in future as the cases of non-zero mass and zero mass respectively.

Angular momentum and spin.

The self adjoint operator corresponding to angular momentum about the x_3 axis is $\frac{1}{i}\dot{T}(m_{12})$. In the case of non-zero mass the little group L_χ is $SU(2)$; in the case of zero mass the little group is Δ . In both cases the 1-parameter subgroup of $SL(2,\mathbb{C})$:

$$t \longmapsto \exp tm_{12} = \begin{pmatrix} e^{\frac{it}{2}} & 0 \\ 0 & e^{\frac{-it}{2}} \end{pmatrix}$$

corresponding to the space rotations of $\mathbb{R}^4$ about the x_3 axis, is a subgroup of L_χ . We can therefore determine $\frac{1}{i}\dot{T}(m_{12})$ by the procedure given at the end of section 7 under the heading: action of H . The assumption that the orbit $G\chi$ has an invariant measure is satisfied in this case, since

$$d\mu(p) = \frac{dp_1 dp_2 dp_3}{p_4}$$

is such a measure, for $\chi = (0,0,0,\sqrt{c})$ or $\chi = (0,0,1,1)$.

The procedure requires the choice of a section $\omega : G\chi \longrightarrow SL(2,\mathbb{C})$ which is equivariant under rotations about the x_3 axis. The section property is

$$\omega(p)\chi = p$$

for all $p \in G\chi$. The physical interpretation of this in the non-zero mass case, is that $\omega(p)$ is a Lorentz transformation to a new frame of reference so that a classical particle with 4-velocity $p = (p_1, p_2, p_3, p_4)$ has new 4-velocity $\chi = (0,0,0,\sqrt{c})$. Thus $\omega(p)$ is a transformation to the rest frame of the particle. In the case of zero mass, $\omega(p)$ is a change to a frame of reference in which a classical photon with 4-velocity p has new 4-velocity $\chi = (0,0,1,1)$ and is therefore moving along the x_3 axis.

A specific equivariant section ω is suggested in each case by the following physical considerations.

i) Non-zero mass.

A convenient section $\omega: G\chi \longrightarrow SL(2,\mathbb{C})$ is defined by taking $\omega(p)$ to be the unique pure Lorentz transformation such $\omega(p)\chi = p$. Physically $\omega(p)$ is a change to an observer moving with uniform velocity. To express this analytically we note that each matrix $h \in SL(2,\mathbb{C})$ has a unique polar decomposition

$$h = \pi u$$

where π is positive definite hermitian and $u \in SU(2)$. This corresponds to the unique expression of a Lorentz transformation as a product of a pure Lorentz transformation and a space rotation; see [36] page 168. Thus $\omega(h\, SU(2)) = \pi$. It follows that for each $v \in SU(2)$ the diagram

$$
\begin{array}{ccc}
SL(2,\mathbf{C}) & \longrightarrow & SL(2,\mathbf{C}) \\
\omega \uparrow & & \uparrow \omega \\
G\chi & \longrightarrow & G\chi
\end{array}
\qquad , \qquad
\begin{array}{ccc}
\pi & \longmapsto & v\pi v^{-1} \\
\uparrow & & \uparrow \\
h\,SU(2) & \longmapsto & vh\,SU(2) \\
\| & & \| \\
\pi\,SU(2) & & v\pi v^{-1}\,SU(2)
\end{array}
$$

is commutative. This shows that ω is $SU(2)$-equivariant.

ii) <u>Zero mass.</u>

In this case we can define $\omega: G\chi \longrightarrow SL(2,\mathbf{C})$ so that $\omega(p)$ is the composition of the pure Lorentz transformation mapping $(0,0,1,1) \longmapsto (0,0,p_4,p_4)$ and the space rotation mapping $(0,0,p_4,p_4) \longmapsto (p_1,p_2,p_3,p_4)$. This section is equivariant under rotations about the x_4 axis. Physically it corresponds to a change of observer given by a change of velocity followed by a rotation.

We can now apply the results of pages 52-55. The section ω reduces the Hilbert space $\mathcal{H}$ of sections of the bundle ξ_σ^χ to a Hilbert space $\mathcal{H}_\omega$ of <u>wave functions</u> defined on the orbit $G\chi$ with values in the representation space V of σ . The Hilbert space V is the fibre of the bundle ξ_σ^χ and the dimension of V is called the number of <u>polarisation states</u> of the relativistic particle. The representation $T = T(\xi_\sigma^\chi)$ defines a representation T_ω on $\mathcal{H}_\omega$. The self-adjoint operator $\frac{1}{i}\dot{T}_\omega(m_{12})$ on $\mathcal{H}_\omega$ representing angular momentum about the x_3 axis can be written as a sum

$$S_\omega^{12} + O_\omega^{12}$$

of self-adjoint operators, where S_ω^{12} has the same spectrum as the operator $\frac{1}{i}\dot{\sigma}(m_{12})$ on V, and where

$$(O_\omega^{12}\psi_\omega)(p) = \frac{1}{i}\frac{d}{dt}\psi_\omega(\exp(-tm_{12})\,p)\,\Big|_{t=0}$$

$$= \frac{1}{i}\frac{d}{dt}\psi_\omega(p_1\cos t - p_2\sin t\,,\ p_1\sin t + p_2\cos t\,,\ p_3,\ p_4)\Big|_{t=0}$$

$$= \frac{1}{i}(p_1\cdot\frac{\partial}{\partial p_2} - p_2\cdot\frac{\partial}{\partial p_1})\,\psi_\omega(p)\,.$$

The physical quantity represented by the operator S_ω^{12} is called the <u>spin angular momentum</u> about the x_3 axis relative to the section ω. The quantity represented by

$$O_\omega^{12} = \frac{1}{i}\,(p_1\cdot\frac{\partial}{\partial p_2} - p_2\cdot\frac{\partial}{\partial p_1})$$

on $\mathcal{H}_\omega$ is called the <u>orbital angular momentum</u> about the x_3 axis relative to the section ω.

We note that, in the non-zero mass case, we have spin and orbital angular momentum defined about all space axes, relative to the same section ω, since ω is SU(2)-equivariant.

The largest absolute value of the eigenvalues of $\frac{1}{i}\dot{\sigma}(m_{12})$ is called the <u>spin</u> of the particle (about the x_3 axis).

Determination of spin.

We will determine the spin of a particle associated with the representation $T(\xi_\sigma^\chi)$. Here σ is an irreducible representation of the little group L_χ and the spin about the x_3 axis is the largest absolute value of the eigenvalues of $\frac{1}{i}\dot{\sigma}(m_{12})$.

i) Non-zero mass.

The little group SU(2) is compact, so that its irreducible representations are finite dimensional, [16] (22.13). They can be described as follows. For each $A \in SU(2)$ let

$$\otimes^r A : \otimes^r C^2 \longrightarrow \otimes^r C^2$$

be its rth tensor power. Let

$$\sigma_r(A): V_{r+1} \longrightarrow V_{r+1}$$

be the restriction of $\otimes^r A$ to the $(r+1)$-dimensional subspace of symmetric tensors. Since SU(2) is compact, [16] (22.23) V_{r+1} can be given a Hilbert space structure so that

$$\sigma_r : SU(2) \longrightarrow U(V_{r+1})$$

is a (unitary) representation. The representations σ_r, $r = 0,1,2,\ldots$ are all irreducible, and every irreducible representation of SU(2) is equivalent to some σ_r.

The element m_{12} in the Lie algebra of $SU(2)$ corresponding to rotations about the x_3 axis is

$$m_{12} = \begin{pmatrix} \frac{1}{2}i & 0 \\ 0 & -\frac{1}{2}i \end{pmatrix} .$$

Therefore $\frac{1}{i}\dot{\sigma}_r(m_{12})$ is a diagonal matrix with entries

$$\frac{r}{2}, \frac{r}{2} - 1, \frac{r}{2} - 2, \ldots, -\frac{r}{2} .$$

Thus the spin about the x_3 axis (and hence about any axis) is $\frac{r}{2}$.

We see that the particle associated with $T(\xi^{\chi}_{\sigma_r})$ has mass $m > 0$ and spin $\frac{r}{2}$ where $\chi = (0,0,0,m^2)$. This proves that the representation associated with a particle of non-zero mass is uniquely determined by its mass $m > 0$ and spin $s = 0, \frac{1}{2}, 1, \frac{3}{2}, \ldots$.

The fibre of the Hilbert bundle associated with a particle of spin $s = \frac{r}{2}$ is V_{r+1} which is $(2s + 1)$-dimensional. Therefore there are $2s + 1$ polarisation states for a particle of spin s and non-zero mass.

ii) <u>Zero mass.</u>

The little group Δ is isomorphic to the semi-direct product

$$\mathbb{C} \circledS \mathbb{R}/_{4\pi}$$

relative to the action $(\theta,z) \longmapsto e^{i\theta}z$ of $\mathbb{R}/_{4\pi}$ on $\mathbb{C}$.

The isomorphism is given by the map

$$\begin{pmatrix} e^{i\theta} & ze^{-i\theta} \\ 0 & e^{-i\theta} \end{pmatrix} \longmapsto (z,2\theta) \ .$$

$\mathbb{C}$ and its dual $\hat{\mathbb{C}}$ can be identified with $\mathbb{R}^2$, so that $\mathbb{R}/_{4\pi}$ acts by rotations. The orbits in $\mathbb{R}^2$ under rotations are the origin and circles with centre the origin. Δ is therefore a regular semi-direct product $\mathbb{R}^2 Ⓢ \mathbb{R}/_{4\pi}$ and we apply the theorem of Mackey to determine its irreducible representations. For each orbit we can choose a representative and determine the little group as follows.

Orbit	Representative	Little group
$\{(0,0)\}$	$(0,0)$	$\mathbb{R}/_{4\pi}$
$\{(x,y) \mid x^2 + y^2 = c^2\}$	$(c,0)$	$\{0,2\pi\}$

No known particles have been associated with representations arising from the orbits $\{(x,y) \mid x^2 + y^2 = c^2\}$. We shall therefore confine ourselves to the orbit $\{(0,0)\}$. Each irreducible representation of $\mathbb{R}/_{4\pi}$ is 1-dimensional and is of the form

$$\theta \longmapsto e^{ri\theta}$$

$r = 0, \pm\frac{1}{2}, \pm 1, \pm\frac{3}{2}, \ldots$, The induced representation of $\mathbb{R}^2 Ⓢ \mathbb{R}/_{4\pi}$ is

$$\tau_r : (x,y,\theta) \longmapsto e^{ri\theta}$$

so that the representation τ_r of Δ is

$$\begin{pmatrix} e^{i\theta} & z \\ 0 & e^{-i\theta} \end{pmatrix} \longmapsto e^{2ri\theta}$$

Thus

$$\exp tm_{12} = \begin{pmatrix} e^{\frac{it}{2}} & 0 \\ 0 & e^{\frac{-it}{2}} \end{pmatrix} \longmapsto e^{rit}$$

so that

$$\frac{1}{i}\,\dot{\tau}_r(m_{12}) = r\ .$$

We see that the particle $T(\xi^X_{\tau_r})$ with mass zero has spin $|r|$, (about the x_3 axis). Moreover, for a given spin s there are exactly two mass zero particles, corresponding to τ_s and τ_{-s}. The Hilbert bundle has a 1-dimensional fibre in each case, so that there are two polarisation states for each spin s and mass zero. The sign of r is called the helicity of the particle.

We conclude with three examples of elementary relativistic free particles. The electron is a particle with non-zero mass and spin $\frac{1}{2}$; the neutrino has mass zero and spin $\frac{1}{2}$; the photon has mass zero and spin 1.

Section 9. The Dirac Equation.

In this section we shall give a brief treatment of the Dirac wave equation, and show that it is associated with a particle of spin $\frac{1}{2}$ and mass $m > 0$.

Minkowski-Clifford algebra.

For each $x \in \mathbb{R}^4$ let $\underset{\sim}{x} = x_1\tau_1 + x_2\tau_2 + x_3\tau_3 + x_4\tau_4$, $\tilde{\underset{\sim}{x}} = -x_1\tau_1 - x_2\tau_2 - x_3\tau_3 + x_4\tau_4$ where the τ_j are the matrices defined in section 3 . Let

$$\gamma(x) = \begin{pmatrix} 0 & \underset{\sim}{x} \\ \tilde{\underset{\sim}{x}} & 0 \end{pmatrix} .$$

Then $\gamma\colon \mathbb{R}^4 \longrightarrow \mathrm{Hom}(\mathbb{C}^4)$ is linear injective and

$$[\gamma(x)]^2 = \langle x,x\rangle 1 .$$

Thus $\gamma(\mathbb{R}^4)$ generates a Clifford algebra over $\mathbb{R}^4$ with respect to the Lorentz scalar product.

The map

$$\tau\colon A \longmapsto \begin{pmatrix} A & 0 \\ 0 & (A^*)^{-1} \end{pmatrix}$$

is an isomorphism of SL(2,$\mathbb{C}$) onto the spin group Spin (3,1) of the Clifford algebra, which is a subgroup of GL(4,$\mathbb{C}$) .
Let $SL(2,\mathbb{C}) \xrightarrow{\sigma} SO(3,1)$ be the covering map, then we have a commutative diagram

$$\begin{array}{ccc} SU(2) & \xrightarrow{\sigma} & SO(3) \\ \downarrow & & \downarrow \\ SL(2,\mathbb{C}) & \xrightarrow{\sigma} & SO(3,1) \\ \tau\downarrow & & \downarrow\gamma_* \\ Spin(3,1) & \longrightarrow & SO(\gamma\mathbb{R}^4) \end{array} .$$

This means simply that

$$\gamma(Ax) = \tau(A)\gamma(x)\tau(A)^{-1}$$

for each $x \in \mathbb{R}^4$ and $A \in SL(2,\mathbb{C})$.

<u>Dirac bundle.</u>

Let $\chi = (0,0,0,m)$ with $m > 0$. Let ξ^{χ}_{τ} be the Hilbert bundle associated with the restriction of the representation τ to SU(2) :

$$\tau(A) = \begin{pmatrix} A & 0 \\ 0 & A \end{pmatrix} \in U(4)$$

all $A \in SU(2)$. Thus ξ^{χ}_{τ} is a bundle with fibre $\mathbb{C}^4$ called the <u>Dirac bundle.</u>

Let $\xi^{\chi}_{\sigma_1}$ be the bundle with fibre $\mathbb{C}^2$ defined by the inclusion representation $SU(2) \xrightarrow{\sigma_1} U(2)$. If we embed $\mathbb{C}^2$ in $\mathbb{C}^4$ by the map

$$(x_1,x_2) \longmapsto (x_1,x_2,x_1,x_2)$$

then $\xi^{\chi}_{\sigma_1}$ is a sub-bundle of ξ^{χ}_{τ}.

The map $G \times_{G_\chi} \mathbb{C}^4 \longrightarrow G\chi \times \mathbb{C}^4$ given by

$$[A,v] \longmapsto (A\chi,\tau(A)v)$$

for each $A \in SL(2,\mathbb{C})$ gives a bundle isomorphism of ξ^{χ}_{τ} onto the product bundle over $G\chi$ with fibre $\mathbb{C}^4$. The Hilbert space structure in the fibres is not preserved however since $SL(2,\mathbb{C}) \xrightarrow{\tau} Spin(3,1)$ is not a unitary representation. By G we mean the group $\mathbb{R}^4 \circledS SL(2,\mathbb{C})$.

Define an action of G on the product bundle by

$$(p,v) \longmapsto (p,x(p)v)$$

for each $x \in \mathbb{R}^4$ and

$$(p,v) \longmapsto (Ap,\tau(A)v)$$

for each $A \in SL(2,\mathbb{C})$. The bundle isomorphism is then a G-isomorphism.

Each section ψ of the bundle ξ^{χ}_{τ} corresponds to a section $p \longmapsto (p,\tilde{\psi}(p))$ of the product bundle, where $\tilde{\psi}$ is a function defined on the orbit $G\chi$ with values in $\mathbb{C}^4$. More specifically, if $\psi(p) = [\omega(p),\varphi(p)]$ where ω is a section $G\chi \longrightarrow SL(2,\mathbb{C})$, then

$$\tilde{\psi}(p) = \tau(\omega(p))\varphi(p) .$$

The function $\tilde{\psi}$ is a solution of the Dirac wave equation:

$$\gamma(p)\tilde{\psi}(p) = m\tilde{\psi}(p)$$

all $p \in G\chi$. This is equivalent to

$$\gamma(p)\tau(\omega(p))\varphi(p) = m\tau(\omega(p))\varphi(p)$$

i.e.

$$\tau(\omega(p))^{-1}\gamma(p)\tau(\omega(p))\varphi(p) = m\varphi(p)$$

i.e.

$$\gamma(\omega(p)^{-1}p)\varphi(p) = m\varphi(p)$$

i.e.

$$\gamma(\chi)\varphi(p) = m\varphi(p)$$

i.e.

$$\begin{pmatrix} 0 & \begin{matrix} m & 0 \\ 0 & m \end{matrix} \\ \begin{matrix} m & 0 \\ 0 & m \end{matrix} & 0 \end{pmatrix} \cdot \varphi(p) = \begin{pmatrix} m & & & 0 \\ & m & & \\ & & m & \\ 0 & & & m \end{pmatrix} \cdot \varphi(p)$$

i.e.

$$\varphi(p) \in \mathbb{C}^2 .$$

Thus $\tilde{\psi}$ satisfies the Dirac equation if and only if ψ is a section of the sub-bundle $\xi^{\chi}_{\sigma_1}$. The particle associated with the bundle $\xi^{\chi}_{\sigma_1}$ has mass m and spin $\frac{1}{2}$, so that the Dirac equation describes such a particle.

Section 10. SU(3) : Charge and Isospin.

We have seen that a representation of the Lie group $\mathbb{R}^4 \circledS SL(2,\mathbb{C})$ on the Hilbert space associated with a quantum mechanical system leads to a definition of the physical concepts of linear momentum, energy, angular momentum, mass, and spin, and leads to a classification of elementary relativistic systems.

More recently the Lie group SU(3) has been used in an effort to explain the quantities electric charge and isospin. We will sketch some of the ideas involved.

Physical interpretation of SU(3).

The Lie algebra of SU(3) is the 8-dimensional algebra su(3) of 3×3 skew hermitian matrices of trace zero. As a physical interpretation, the matrix

$$Q = \begin{pmatrix} \frac{1}{3}i & & 0 \\ & -\frac{2}{3}i & \\ 0 & & \frac{1}{3}i \end{pmatrix}$$

is associated with **electric charge** and the matrices

$$I_1 = \begin{pmatrix} 0 & \frac{1}{2}i & 0 \\ \frac{1}{2}i & 0 & 0 \\ 0 & 0 & 0 \end{pmatrix} \qquad I_2 = \begin{pmatrix} 0 & \frac{1}{2} & 0 \\ -\frac{1}{2} & 0 & 0 \\ 0 & 0 & 0 \end{pmatrix} \qquad I_3 = \begin{pmatrix} \frac{1}{2}i & 0 & 0 \\ 0 & -\frac{1}{2}i & 0 \\ 0 & 0 & 0 \end{pmatrix}$$

are associated with the three components of <u>isospin</u>.

If it is assumed that, for a given quantum mechanical system with Hilbert space H, we have representations of $G = \mathbb{R}^4 \circledS SL(2,\mathbb{C})$ and $SU(3)$ on H which commute, then we will have a representation T of the direct product $G \times SU(3)$ on H. If the system is an elementary one in the sense that T is irreducible, then T will be of the form $T = \sigma \otimes \tau$ where σ is an irreducible representation of G on a Hilbert space H_o, τ is an irreducible representation of $SU(3)$ on a Hilbert space W, and $H = H_o \otimes W$. Since $SU(3)$ is compact W is finite dimensional.

Let $su(3) \xrightarrow{\dot{\tau}} U(W)$ be the Lie algebra representation defined by τ. This extends to a unique homomorphism of complex Lie algebras

$$sl(3,\mathbb{C}) \xrightarrow{\dot{\tau}} Hom(W)$$

where $sl(3,\mathbb{C})$ is the complex Lie algebra of 3×3 complex matrices with trace zero. Let $\mathcal{C}$ be the abelian (Cartan) sub-algebra of $sl(3,\mathbb{C})$ spanned by Q and I_3:

$$\mathcal{C} = \left\{ c = \begin{pmatrix} c_1 & & 0 \\ & c_2 & \\ 0 & & -(c_1+c_2) \end{pmatrix} \middle| \; c_i \in C \right\}$$

Since Q and I_3 commute, the skew-adjoint operators $\dot{\tau}(Q)$ and $\dot{\tau}(I_3)$ have common eigenvectors

$$e_1, \dots , e_n$$

which form a basis for W. The e_j are eigenvectors of $\dot{\tau}(X)$ for all $X \in \mathcal{L}$; let $\omega_j(X)$ be the eigenvalue of $\dot{\tau}(X)$ on e_j. The linear forms $\omega_1,\dots,\omega_n$ on the complex vector space $\mathcal{L}$ are called **weights** of the representation $\dot{\tau}$. The eigenvectors $e_1,\dots,e_n$ are called **weight vectors** of $\dot{\tau}$.

We have an isomorphism

$$H = H_o \otimes W \approx H_o \oplus \dots \oplus H_o \qquad \text{(n factors)}$$

defined by

$$\phi \otimes e_j \longmapsto (o,\dots,o,\phi,o,\dots,o) .$$

(Moreover this decomposition of H gives a decomposition of the representation T when restricted to G :

$$T|_G = \sigma \oplus \dots \oplus \sigma \quad .$$

Since σ is an irreducible representation of $G = \mathbb{R}^4 \circledS SL(2,\mathbb{C})$ it is associated with a relativistic elementary particle of a definite mass m and spin s say. Thus the system will, under observation, manifest itself as any one of n particles each of mass m and spin s. Furthermore, for $X = Q$ or I_3, $\frac{1}{i}\dot{T}(X)$ has a discrete spectrum

$$\frac{1}{i}\,\omega_1(X) \;, \dots , \frac{1}{i}\,\omega_n(X)$$

and $H = H_0 \oplus \ldots \oplus H_0$ is the corresponding eigenspace decomposition. Thus the jth subspace H_0 represents an elementary relativistic particle of mass m, spin s, electric charge $\frac{1}{i}\omega_j(Q)$ and 3rd component of isospin $\frac{1}{i}\omega_j(I_3)$.

Determination of weights.

It remains to determine the weights ω_j associated with the possible irreducible representations of $SU(3)$, and to identify the corresponding particles.

The weights of the adjoint representation of $sl(3,\mathbb{C})$ are called the roots of $sl(3,\mathbb{C})$ and are

$$0 \, , \, 0 \, , \, \alpha \, , \, \beta \, , \, -\alpha, \, -\beta \, , \, \alpha + \beta, \, -\alpha - \beta$$

where α and β are the linear forms on $\mathcal{C}$:

$$c \longmapsto -c_1 - 2c_2 \quad \text{and} \quad c \longmapsto 2c_1 + c_2 \quad \text{respectively .}$$

Let $\mathcal{C}'$ be the vector space of linear forms on $\mathcal{C}$ (dual of $\mathcal{C}$) and let V be the rational vector space contained in $\mathcal{C}'$ and spanned by α and β. The bilinear (Killing) form on $sl(3,\mathbb{C})$ defined by

$$\langle x,y \rangle = \text{trace } (\text{ad}x \circ \text{ad}y)$$

is a symmetric scalar product whose restriction to $\mathcal{C}$ is non-singular. This induces a symmetric scalar product $\langle \cdot,\cdot \rangle$ on $\mathcal{C}'$. The restriction of this to V is positive definite and rational valued, so that V is a rational Euclidean space when equipped with this scalar product.

The vectors $\frac{1}{i} I_3$ and $\frac{1}{i} M = \frac{1}{i}\sqrt{3}(Q-I_3)$ are orthogonal in $\mathfrak{h}$ and equal in length. Relative to the dual basis in $\mathfrak{h}'$ any element ω in $\mathfrak{h}'$ will have coordinates

$$\left(\frac{1}{i}\,\omega(I_3)\ ,\ \frac{1}{i}\,\omega(M) \right) .$$

In particular the root α has coordinates $(\frac{1}{2}, \frac{\sqrt{3}}{2})$ and β has coordinates $(\frac{1}{2}\ ,\ -\frac{\sqrt{3}}{2})$. By our choice of basis vectors in $\mathfrak{h}$ which are orthogonal and equal in length we ensure that the map $V \longrightarrow \mathbb{R}^2$:

$$\omega \longmapsto \left(\frac{1}{i}\,\omega(I_3)\ ,\ \frac{1}{i}\,\omega(M) \right)$$

preserves angles.

We shall use the following facts about weights, which are all special cases of general theorems on the weights of semi-simple algebras. See [20] IV Theorem 1, VII, [30] LA 7, and [29] .

1 All weights of all finite dimensional representations of $sl(3,\mathbb{C})$ belong to the rational Euclidean space V .

2 If γ is any root and ω is any weight of a finite dimensional representation $\dot{\tau}$, then the reflection of ω (as a point in the Euclidean space V) in the line perpendicular to γ is also a weight of $\dot{\tau}$. The set

$$\{ r \in \mathbb{Z} \mid \omega + r\gamma \text{ is a weight of } \dot{\tau} \}$$

is an unbroken interval in $\mathbb{Z}$.

3 Among the weights ω of an irreducible representation $\dot{\tau}$ there is just one with $\frac{1}{i}\,\omega(I_3)$ maximal. We call this the <u>highest weight</u> of $\dot{\tau}$.

4 All the weights of an irreducible representation $\dot{\tau}$ can be obtained from the highest weight of $\dot{\tau}$ by repeated applications of property 2. The highest weight determines $\dot{\tau}$ up to equivalence.

5 If ω is the highest weight of an irreducible representation $\dot{\tau}$ then

$$p = \frac{2 < \omega,\alpha >}{< \alpha,\alpha >} \qquad \text{and} \quad q = \frac{2 < \omega,\beta >}{< \beta,\beta >}$$

are non-negative integers. For each pair (p,q) of non-negative integers there is a (by 4 unique) irreducible representation $\dot{\tau}(p,q)$ with highest weight ω such that $\frac{2 < \omega,\alpha >}{< \alpha,\alpha >} = p$, $\frac{2 < \omega\,\beta >}{< \beta,\beta >} = q$, so that

$$\omega = \left(\frac{p + q}{2} \, , \, \frac{1}{2\sqrt{3}}\,(p - q) \right),$$ using the coordinates $V \longrightarrow \mathbb{R}^2$

defined above.

<u>Examples</u>.

i) The representation $\dot{\tau}(1,1)$ has highest weight with coordinates $(1,0)$ which is the same as the root $\alpha + \beta$ which is the highest weight of the adjoint representation. Thus $\dot{\tau}(1,1)$ is equivalent to the adjoint representation and the weight diagram is:

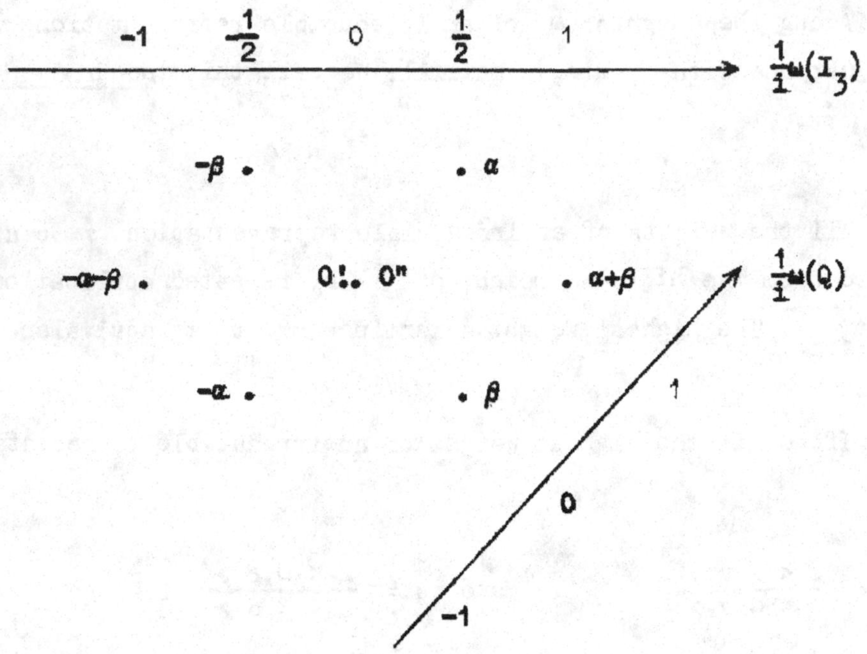

ii) The representation $\dot{\gamma}(3,0)$ has highest weight $(\frac{3}{2}, \frac{\sqrt{3}}{2})$.

By repeated application of property 2 we obtain the weight diagram:

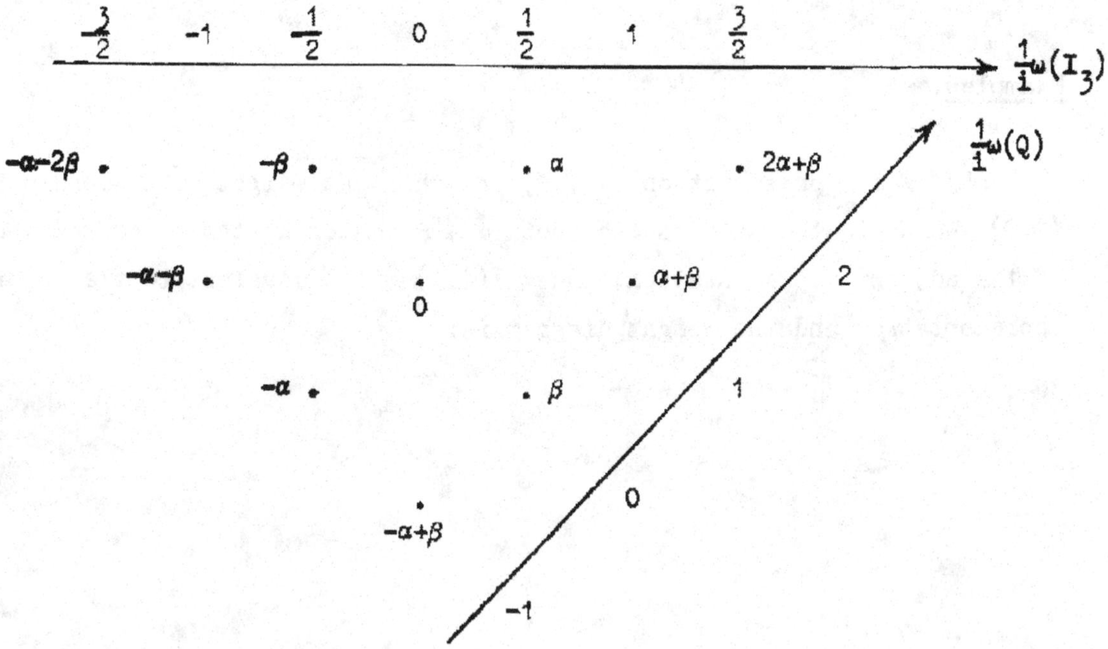

<u>Particle assignments.</u>

According to the theory developed above a quantum mechanical system associated with an irreducible representation of the direct product $(\mathbb{R}^4 \circledS SL(2,\mathbb{C})) \times SU(3)$ will be associated with an irreducible representation of $\mathbb{R}^4 \circledS SL(2,\mathbb{C})$, giving a mass m and a spin s say, and an irreducible representation of $SU(3)$ which gives a set of weights $\omega_1,\ldots,\omega_n$ say. The system then consists of n particles each with the same mass m and same spin s, and the jth particle will have electric charge $\frac{1}{i}\omega_j(Q)$ and 3rd component of isospin $\frac{1}{i}\omega_j(I_3)$

In pratice however the theory is applied to systems of particles whose masses are not equal, although some are of the same order of magnitude. We give some of these systems below.

The first list gives some particles which are associated with the adjoint representation of $SU(3)$. We use the conventional names or symbols for these particles. Masses are given in MEV units. Each particle is listed under the corresponding weight (root in this case).

Weights:	$-\beta$	α	$-\alpha-\beta$	$0'$	$\alpha+\beta$	$-\alpha$	β	$0''$
Representation of $R^4 Ⓢ SL(2,\mathbb{C})$:								
i) baryons mass ≃ 1130 ± 192 spin $\frac{1}{2}$	neutron	proton	Σ^-	Σ^0	Σ^+	Ξ^-	Ξ^0	Λ^0
ii) anti-baryons mass ≃ 1130 ± 192 spin $\frac{1}{2}$	$\bar{\Xi}^0$	$\bar{\Xi}^+$	$\bar{\Sigma}^-$	$\bar{\Sigma}^0$	$\bar{\Sigma}^+$	anti pro- ton	anti neu- tron	$\bar{\Lambda}^0$
iii) pseudo scalar mesons mass ≃ 315 ± 182 spin 0	K^0	K^+	π^-	π^0	π^+	$\bar{K}^-$	$\bar{K}^0$	η
iv) vector mesons mass ≃ 800 spin 1	K^{*0}	K^{*+}	ρ^-	ρ^0	ρ^+	$\bar{K}^{*-}$	$\bar{K}^{*0}$	φ^0
v) 2nd π-meson-nucleon resonance mass ≃ 1600 spin $\frac{3}{2}$	N^{**0}	N^{**+}	Y_1^{-**}	Y_1^{0**}	Y_1^{+**}	Ξ^{-**}	Ξ^{0**}	Y_0^{**}
vi) 3rd π-meson-nucleon resonance mass ≃ 1688 spin $\frac{5}{2}$	N^{***0}	....	....	....	....	....	....	Y_0^{***}

The second list gives some particles associated with the 10 dimensional representation $\dot{\tau}(3,0)$ of SU(3) .

Weights:	$-\alpha-2\beta$	$-\beta$	α	$2\alpha+\beta$	$-\alpha-\beta$	0	$\alpha+\beta$	$-\alpha$	β	$-\alpha+\beta$
Representation of $R^4 \circledS SL(2,\mathbb{C})$:										
i) meson-baryon resonance mass = 1460 ± 223 spin $\frac{3}{2}$	$\overset{-}{N^*}$	$\overset{o}{N^*}$	$\overset{1}{N^*}$	$\overset{2}{N^*}$	$\overset{-}{Y^*}$	$\overset{o}{Y^*}$	$\overset{+}{Y^*}$	$\overset{-}{\Xi^*}$	$\overset{o}{\Xi^*}$	Ω^-
ii) 4th π-meson-nucleon resonance mass = 1922 spin $\frac{7}{2}$	$\overset{-}{N^{****}}$	.....	.....	.....	.....	.....	.....	.....	.....	.....

This information is taken from Gourdin [21] .

Index of Terms.

With page of first occurence.

Bibliography.

[1] V. Bargmann, On unitary ray representations of continuous groups, Ann. Math. 59, 1-46, 1954

[2] V. Bargmann, Note on Wigner's theorem on symmetry operations, Jour. Math. Physics 5, 862-868, 1964

[3] A. Borel, Topology of Lie groups and characteristic classes, Bull. Amer. Math. Soc. 61, 397-432, 1955

[4] N. Bourbaki, Algebre Multilinéaire, Hermann 1948

[5] N. Bourbaki, Mesure de Haar, Hermann 1963

[6] H. Cartan and S. Eilenberg, Homological Algebra, Princeton 1956

[7] C. Chevalley, Theory of Lie Groups Vol. 1, Princeton 1946

[8] C. Chevalley, Théorie des Groupes de Lie Vol. 2, Hermann

[9] C. Chevalley, Théorie des Groupes de Lie, Vol 3, Hermann

[10] C. Chevalley, Fundamental Concepts of Algebra, Academic Press 1956

[11] F. Dyson, Symmetry Groups in Nuclear and Particle Physics, Benjamin 1966

[12] S. Eilenberg and N. Steenrod, Foundations of Algebraic Topology, Princeton 1952

[13] I. Gelfand, R. Minlos, Z. Shapiro, Representations of the Rotation and Lorentz Groups, Pergamon 1963

[14] M. Gell-Mann, The Eight-Fold Way, Benjamin 1964

[15] R. Hermann, Lie Groups for Physicists, Benjamin 1966

[16] E. Hewitt, K. Ross, Abstract Harmonic Analysis Vol 1, Springer 1963

[17] F. Hirzebruch, Topological Methods in Algebraic Geometry, Springer 1966

[18] G. Hochschild, Structure of Lie Groups, Holden Day 1966

[19] S.-T- Hu, Homotopy Theory, Academic Press 1959

[2o] N. Jacobson, Lie Algebras, Intersience 1962

[21] M. Gourdin, Some topics related to unitary symmetry, in Springer Tracts in Modern Physics 36.

[22] G.W. Mackey, Induced representations of locally compact groups I, Ann. Math. 55, 1o1-139, 1952

[23] G.W. Mackey, Imprimitivity for representations of locally compact groups I, Proc. Nat. Acad. Sci. 35, 537-545, 1949

[24] G.W. Mackey, The Theory of Group Representations, Univ. of Chicago lecture notes, 1955

[25] G.W. Mackey, Group Representations and Non-commútative Harmonic Analysis, Berkeley lecture notes, 1965

[26] G.W. Mackey, Mathematical Foundations of Quantum Mechanics, Benjamin 1963

[27] D. Montgomery and L. Zippin, Topological Transformation Groups, Intersience 1955

[28] M.A. Naimark, Normed Rings, Noordhof 1964

[29] G. Racah, Group theory and spectroscopy, Springer Tracts in Modern Physics 37

[3o] J.P. Serre, Lie Algebras and Lie Groups, Benjamin 1965

[31] I. Segal, A class of operator algebras, Duke. Math. J. 18, 221-265, 1951

[32] I. Segal, Mathematical Problems of Relativistic Physics, Am. Math. Soc. 1963

[33] E. Spanier, Algebraic Topology, McGraw-Hill 1966

[34] M.H. Stone, On one-parameter unitary groups in Hilbert space, Ann. Math. 33, 643-648, 1932

[35] N. Steenrod, The Topology of Fibre Bundles, Princeton 1951

[36] A.S. Wightman, L'invariance dans la mécanique quantique relativiste in Relations de Dispersion et particules Élémentaires, Hermann 196o

[37] E.P. Wigner, On unitary representations of the inhomogeneous Lorentz group, Ann. of Math. 4o, 149-2o4, 1939

[38] E.P. Wigner, Group Theory, Academic Press 1959

[39] E. Wigner and V. Bargmann, Group theoretical discussion of relativistic wave equations, Proc. Nat. Acad. Sci. 34, 211-223, 1948

[4o] E.C. Zeeman, Causality implies the Lorentz group, Jour. Math. Physics 5, 49o-493, 1964

Offsetdruck: Julius Beltz, Weinheim/Bergstr.

Lecture Notes in Mathematics

Bisher erschienen/Already published

Vol. 1: J. Wermer, Seminar über Funktionen-Algebren. IV, 30 Seiten. 1964. DM 3,80 / $ 0.95

Vol. 2: A. Borel, Cohomologie des espaces localement compacts d'après J. Leray. IV, 93 pages. 1964. DM 9,– / $ 2.25

Vol. 3: J. F. Adams, Stable Homotopy Theory. 2nd. revised edition. IV, 78 pages. 1966. DM 7,80 / $ 1.95

Vol. 4: M. Arkowitz and C. R. Curjel, Groups of Homotopy Classes. 2nd. revised edition. IV, 36 pages. 1967. DM 4,80 / $ 1.20

Vol. 5: J.-P. Serre, Cohomologie Galoisienne. Troisième édition. VIII, 214 pages. 1965. DM 18,– / $ 4.50

Vol. 6: H. Hermes, Eine Termlogik mit Auswahloperator. IV, 42 Seiten. 1965. DM 5,80 / $ 1.45

Vol. 7: Ph. Tondeur, Introduction to Lie Groups and Transformation Groups. VIII, 176 pages. 1965. DM 13,50 / $ 3.40

Vol. 8: G. Fichera, Linear Elliptic Differential Systems and Eigenvalue Problems. IV, 176 pages. 1965. DM 13.50 / $ 3.40

Vol. 9: P. L. Ivănescu, Pseudo-Boolean Programming and Applications. IV, 50 pages. 1965. DM 4,80 / $ 1.20

Vol. 10: H. Lüneburg, Die Suzukigruppen und ihre Geometrien. VI, 111 Seiten. 1965. DM 8,– / $ 2.00

Vol. 11: J.-P. Serre, Algèbre Locale. Multiplicités. Rédigé par P. Gabriel. Seconde édition. VIII, 192 pages. 1965. DM 12,– / $ 3.00

Vol. 12: A. Dold, Halbexakte Homotopiefunktoren. II, 157 Seiten. 1966. DM 12,– / $ 3.00

Vol. 13: E. Thomas, Seminar on Fiber Spaces. IV, 45 pages. 1966. DM 4,80 / $ 1.20

Vol. 14: H. Werner, Vorlesung über Approximationstheorie. IV, 184 Seiten und 12 Seiten Anhang. 1966. DM 14,– / $ 3.50

Vol. 15: F. Oort, Commutative Group Schemes. VI, 133 pages. 1966. DM 9,80 / $ 2.45

Vol. 16: J. Pfanzagl and W. Pierlo, Compact Systems of Sets. IV, 48 pages. 1966. DM 5,80 / $ 1.45

Vol. 17: C. Müller, Spherical Harmonics. IV, 46 pages. 1966. DM 5,– / $ 1.25

Vol. 18: H.-B. Brinkmann und D. Puppe, Kategorien und Funktoren. XII, 107 Seiten. 1966. DM 8,– / $ 2.00

Vol. 19: G. Stolzenberg, Volumes, Limits and Extensions of Analytic Varieties. IV, 45 pages. 1966. DM 5,40 / $ 1.35

Vol. 20: R. Hartshorne, Residues and Duality. VIII, 423 pages. 1966. DM 20,– / $ 5.00

Vol. 21: Seminar on Complex Multiplication. By A. Borel, S. Chowla, C. S. Herz, K. Iwasawa, J-P. Serre. IV, 102 pages. 1966. DM 8,– / $ 2.00

Vol. 22: H. Bauer, Harmonische Räume und ihre Potentialtheorie. IV, 175 Seiten. 1966. DM 14,– / $ 3.50

Vol. 23: P. L. Ivănescu and S. Rudeanu, Pseudo-Boolean Methods for Bivalent Programming. 120 pages. 1966. DM 10,– / $ 2.50

Vol. 24: J. Lambek, Completions of Categories. IV, 69 pages. 1966. DM 6,80 / $ 1.70

Vol. 25: R. Narasimhan, Introduction to the Theory of Analytic Spaces. IV, 143 pages. 1966. DM 10,– / $ 2.50

Vol. 26: P.-A. Meyer, Processus de Markov. IV, 190 pages. 1967. DM 15,– / $ 3.75

Vol. 27: H. P. Künzi und S. T. Tan, Lineare Optimierung großer Systeme. VI, 121 Seiten. 1966. DM 12,– / $ 3.00

Vol. 28: P. E. Conner and E. E. Floyd, The Relation of Cobordism to K-Theories. VIII, 112 pages. 1966. DM 9.80 / $ 2.45

Vol. 29: K. Chandrasekharan, Einführung in die Analytische Zahlentheorie. VI, 199 Seiten. 1966. DM 16.80 / $ 4.20

Vol. 30: A. Frölicher and W. Bucher, Calculus in Vector Spaces without Norm. X, 146 pages. 1966. DM 12,– / $ 3.00

Bitte wenden / Continued

Vol. 31: Symposium on Probability Methods in Analysis. Chairman: D.A.Kappos. IV, 329 pages. 1967. DM 20,– / $ 5.00

Vol. 32: M. André, Méthode Simpliciale en Algèbre Homologique et Algèbre Commutative. IV, 122 pages. 1967. DM 12,– / $ 3.00

Vol. 33: G. I. Targonski, Seminar on Functional Operators and Equations. IV, 110 pages. 1967. DM 10,– / $ 2.50

Vol. 34: G. E. Bredon, Equivariant Cohomology Theories. VI, 64 pages. 1967. DM 6,80 / $ 1.70

Vol. 35: N. P. Bhatia and G. P. Szegö, Dynamical Systems: Stability Theory and Applications. VI, 416 pages. 1967. DM 24,– / $ 6.00

Vol. 36: A. Borel, Topics in the Homology Theory of Fibre Bundles. VI, 95 pages. 1967. DM 9,– / $ 2.25

Vol. 37: R. B. Jensen, Modelle der Mengenlehre. X, 176 Seiten. 1967. DM 14,– / $ 3.50

Vol. 38: R. Berger, R. Kiehl, E. Kunz und H.-J. Nastold, Differentialrechnung in der analytischen Geometrie. IV, 134 Seiten. 1967. DM 12,– / $ 3.00

Vol. 39: Séminaire de Probabilités I. II, 189 pages. 1967. DM 14,– / $ 3.50

Vol. 40: J. Tits, Tabellen zu den einfachen Lie Gruppen und ihren Darstellungen. VI, 53 Seiten. 1967. DM 6,80 / $ 1.70

Vol. 41: R. Hartshorne, Local Cohomology. VI, 106 pages. 1967. DM 10,– / $ 2.50

Vol. 42: J. F. Berglund and K. H. Hofmann, Compact Semitopological Semigroups and Weakly Almost Periodic Functions. VI, 160 pages. 1967. DM 12,– / $ 3.00

Vol. 43: D. G. Quillen, Homotopical Algebra. VI, 157 pages. 1967. DM 14,– / $ 3.50

Vol. 44: K. Urbanik, Lectures on Prediction Theory. IV, 50 pages. 1967. DM 5,80 / $ 1.45

Vol. 45: A. Wilansky, Topics in Functional Analysis. VI, 102 pages. 1967. DM 9,60 / $ 2.40

Vol. 46: P. E. Conner, Seminar on Periodic Maps. IV, 116 pages. 1967. DM 10,60 / $ 2.65

Vol. 47: Reports of the Midwest Category Seminar. IV, 181 pages. 1967. DM 14,80 / $ 3.70

Vol. 48: G. de Rham, S. Maumary and M. A. Kervaire, Torsion et Type Simple d'Homotopie. IV, 101 pages. 1967 DM 9,60 / $ 2.40

Vol. 49: C. Faith, Lectures on Injective Modules and Quotient Rings. XVI, 140 pages. 1967. DM 12,80 / $ 3.20

Vol. 50: L. Zalcman, Analytic Capacity and Rational Approximation. VI, 155 pages. 1968. DM 13,20/$ 3.40

Vol. 51: Séminaire de Probabilités II. IV, 199 pages. 1968. DM 14,– / $ 3.50

Beschaffenheit der Manuskripte

Die Manuskripte werden photomechanisch vervielfältigt; sie müssen daher in sauberer Schreibmaschinenschrift geschrieben sein. Handschriftliche Formeln bitte nur mit schwarzer Tusche oder roter Tinte eintragen. Korrekturwünsche werden in der gleichen Maschinenschrift auf einem besonderen Blatt erbeten (Zuordnung der Korrekturen im Text und auf dem Blatt sind durch Bleistiftziffern zu kennzeichnen). Der Verlag sorgt dann für das ordnungsgemäße Tektieren der Korrekturen. Falls das Manuskript oder Teile desselben neu geschrieben werden müssen, ist der Verlag bereit, dem Autor bei Erscheinen seines Bandes einen angemessenen Betrag zu zahlen. Die Autoren erhalten 25 Freiexemplare.

Manuskripte, in englischer, deutscher oder französischer Sprache abgefaßt, nimmt Prof. Dr. A. Dold, Mathematisches Institut der Universität Heidelberg, Tiergartenstraße oder Prof. Dr. B. Eckmann, Eidgenössische Technische Hochschule, Zürich, entgegen.

Cette série a pour but de donner des informations rapides, de niveau élevé, sur des développements récents en mathématiques, aussi bien dans la recherche que dans l'enseignement supérieur. On prévoit de publier

1. des versions préliminaires de travaux originaux et de monographies
2. des cours spéciaux portant sur un domaine nouveau ou sur des aspects nouveaux de domaines classiques
3. des rapports de séminaires
4. des conférences faites à des congrès ou des colloquiums

En outre il est prévu de publier dans cette série, si la demande le justifie, des rapports de séminaires et des cours multicopiés ailleurs et qui sont épuisés.

Dans l'intéret d'une diffusion rapide, les contributions auront souvent un caractère provisoire; le cas échéant, les démonstrations ne seront données que dans les grandes lignes, et les résultats et méthodes pourront également paraître ailleurs. Par cette série de »prépublications« les éditeurs Springer espèrent rendre d'appréciables services aux instituts de mathématiques par le fait qu'une réserve suffisante d'exemplaires sera toujours disponibles et que les personnes intéressées pourront plus facilement être atteintes. Les annonces dans les revues spécialisées, les inscriptions aux catalogues et les copyrights faciliteront pour les bibliothèques mathématiques la tâche de réunir une documentation complète.

Présentation des manuscrits

Les manuscrits, étant reproduits par procédé photomécanique, doivent être soigneusement dactylographiés. Il est demandé d'écrire à l'encre de Chine ou à l'encre rouge les formules non dactylographiées. Des corrections peuvent également être dactylographiées sur une feuille séparée (prière d'indiquer au crayon leur ordre de classement dans le texte et sur la feuille), la maison d'édition se chargeant ensuite de les insérer à leur place dans le texte. S'il s'àvère nécessaire d'écrire de nouveau le manuscrit, soit complètement, soit en partie, la maison d'édition se déclare prête à se charger des frais à la parution du volume. Les auteurs recoivent 25 exemplaires gratuits.

Les manuscrits en anglais, allemand ou français peuvent être adressés au Prof. Dr. A. Dold, Mathematisches Institut der Universität Heidelberg, Tiergartenstraße ou au Prof. Dr. B. Eckmann, Eidgenössische Technische Hochschule, Zürich.

Printed in the United States
by Bookmasters

Printed in the United States
By Bookmasters